Organic Elemental Analysis

Ultramicro, Micro, and Trace Methods

Organic Elemental Analysis

Ultramicro, Micro, and Trace Methods

WOLFGANG J. KIRSTEN
The National Swedish Laboratory for Agricultural Chemistry
Uppsala, Sweden

1983

ACADEMIC PRESS
A Subsidiary of Harcourt Brace Jovanovich, Publishers
New York London
Paris San Diego San Francisco São Paulo Sydney Tokyo Toronto

ACADEMIC PRESS, INC.
111 Fifth Avenue, New York, New York 10003

United Kingdom Edition published by
ACADEMIC PRESS, INC. (LONDON) LTD.
24/28 Oval Road, London NW1 7DX

Library of Congress Cataloging in Publication Data

Kirsten, Wolfgang J.
Organic elemental analysis.

Bibliography: p.
Includes index.
1. Chemistry, Analytic. 2. Chemistry, Organic.
I. Title.
QD271.K57 1983 547.3'483 83-2805
ISBN 0-12-410280-8

PRINTED IN THE UNITED STATES OF AMERICA

83 84 85 86 9 8 7 6 5 4 3 2 1

CONTENTS

Preface ix
Acknowledgments xi

INTRODUCTION 1

Chapter 1 REAGENT SOLUTIONS, STANDARD SOLUTIONS, AND STANDARD SUBSTANCES 6

Chapter 2 TRANSFERS 11

Chapter 3 HOMOGENIZATION OF SPECIMENS 15

Chapter 4 DRYING OF SPECIMENS AND SAMPLES 17

Chapter 5 WEIGHING AND ENCAPSULATION OF SAMPLES

Weighing in Crimped Capsules 22
Hygroscopic Samples 23
Volatile and Hygroscopic Liquids 26
Volatile Solids, Smeary and Oily Materials, and Wet Solids 27
Precautions 29

Chapter 6 **SPECTROPHOTOMETRY** 31

Chapter 7 **MULTIELEMENT DETERMINATIONS** 33

Chapter 8 **THE CARLO ERBA AUTOMATIC GAS CHROMATOGRAPHIC ELEMENTAL ANALYZER**

The Function of the CHN Channel 38
The Function of the Oxygen Channel 39
Processing of the Data 39
Some Common Operations 41
General Remarks 41

Chapter 9 **AUTOMATIC SIMULTANEOUS DETERMINATION OF CARBON, HYDROGEN, AND NITROGEN**

Apparatus 43
Reagents 43
Adjustment of Apparatus 46
Procedure 50
Applicability 51

Chapter 10 **AUTOMATIC SIMULTANEOUS DETERMINATION OF CARBON, HYDROGEN, NITROGEN, AND SULFUR**

Apparatus 55
Reagents 55
Adjustment of Apparatus 57
Procedure 58
Applicability 60

Chapter 11 **AUTOMATIC SIMULTANEOUS DETERMINATION OF CARBON, HYDROGEN, NITROGEN, AND SULFUR OR TRACE SULFUR**

Apparatus 62
Reagents 63
Adjustment of Apparatus 65
Procedure 65

Chapter 12 **AUTOMATIC MICRO AND TRACE DETERMINATION OF NITROGEN**

Apparatus 69
Modifications 69
Reagents 73
Adjustment of Apparatus 74
Encapsulation of Samples and Loading of Drums 74
Running the Analyses 76
Applicability 77

Chapter 13 **ADDITIONAL REMARKS ON CHN, CHNS, CHNS TRACE S, AND N DETERMINATION METHODS**

Reactor Tubes 79
Retention of Nitrogen 80
Inlet Delay Time 81
Temperature of Copper Filling 81

Chapter 14 **AUTOMATIC DETERMINATION OF OXYGEN**

Apparatus 83
Reagents 85
Adjustment of Apparatus 86
Procedure 87
Applicability 88
Discussion of the Oxygen Determination Method 89

Chapter 15 **ULTRAMICRO DETERMINATION OF SULFUR**

Reagents 92
Additional Reagents for Extraction Method 94
Additional Reagent for Bismuth-, Iodine-, Mercury-, or Selenium-Containing Samples 94
Assembly and Adjustment of Apparatus 95
Procedure 95
Discussion 99

Chapter 16 **ULTRAMICRO DETERMINATION OF FLUORINE**

Reagents 102
Assembly and Adjustment of Apparatus 103
Procedures 103

Chapter 17 **SIMULTANEOUS DETERMINATION OF SULFUR AND FLUORINE**

Apparatus 105
Procedure 106

Chapter 18 **TRACE DETERMINATION OF SULFUR OR FLUORINE**

Apparatus 108
Additional Reagents for Fluorine (Seldom Needed) 108
Adjustment of Apparatus 109
Combustion Methods 110

Chapter 19 **ULTRAMICRO DETERMINATION OF CHLORINE, BROMINE, AND IODINE**

Hot-Flask Combustion 115
Titration 119
Precautions 120

Chapter 20 **TRACE DETERMINATION OF CHLORINE, BROMINE, AND IODINE**

Apparatus 121
Reagents 121
Procedure 123

Chapter 21 **ULTRAMICRO AND TRACE DETERMINATION OF PHOSPHORUS AND PHOSPHATE**

Reagents 127
Procedure: Ultramicro and Micro 131
Procedure: Trace 131
Spectrophotometry 132
Interferences 133

Chapter 22 **DETERMINATION OF ASHES** 135

REFERENCES 139

Index 143

PREFACE

This book contains a collection of methods for ultramicro, micro, and trace organic elemental analysis that are used in the author's laboratory for commercial routine analysis. Many of the samples analyzed in this laboratory are research samples, but the overwhelming number consist of technical, industrial, and agricultural raw materials and intermediate and final products. The methods described have been found to compare favorably with regard to speed, accuracy, reliability, and economy not only with other methods used in research work, but also with the macro and semimicro methods commonly employed in industrial and agricultural laboratories. Previously, such laboratories had hardly ever analyzed such materials with ultramicro and micro methods, both because it was too difficult to obtain representative samples and because the methods were not suited to a commercial type of analysis.

The recent development of very fast, accurate, and automated instruments for single-element and multielement determinations, the availability of fast electronic ultramicro balances which are sturdy and can be placed in almost any laboratory without special precautions, and the development of efficient and fast homogenization methods have changed this situation. Microgram methods are very often faster and more accurate than the conventional macro methods, and the use of smaller samples and smaller amounts of reagents greatly improves the working conditions in the laboratory.

It is, therefore, the author's hope, that not only the research analyst, but also the industrial analyst, who is responsible for the fast and economic production of large numbers of analyses, will find this book useful.

ACKNOWLEDGMENTS

The author is greatly indebted to the late Einar Stenhagen, who initiated the establishment of the author's laboratory for organic micro and ultramicro elemental analysis in 1943 and who promoted its development over the years.

The importance of the experiences that have been obtained through contact with the pioneering work of such scientists as the late Paul L. Kirk in Berkeley, the late Ronald Belcher in Birmingham, Walter Walisch in Saarbrücken, Günthur Tölg in Mainz, and Franco Poy and Ermes Pella in Milan is gratefully acknowledged.

The author is very indebted to the late Eric Swanbeck, Nicroma, Stockholm, for his very helpful cooperation with the design and construction of high-temperature furnaces and other apparatus.

Gunnar U. Hesselius has led the commercial section of the laboratory since 1971. He has contributed to the development of methods through his own experimental work and by valuable suggestions and discussions.

Economic support of the methods development work was obtained from the Swedish Medical Research Council, the Swedish Natural Science Research Council, the Swedish Technical Research Council, the Knut and Alice Wallenberg Foundation, and later from the Swedish Council for Forestry and Agricultural Research, the Faculty of Agriculture of the Swedish University of Agricultural Sciences, and the methods development fund of the Swedish National Laboratory for Agricultural Chemistry.

INTRODUCTION

Since the foundation of the author's microanalytical laboratory in 1943 many analytical methods have been used. Many of them have been modified, and many have been discarded when better methods became available. The methods described in this book are those that are routinely used now. Some of the more important features of the methods are listed in Table I.

Dynamic methods, that is, methods in which the decomposition of the sample, the chemical reactions, and the measurement are carried out in a continuous flow, have great advantages. They are very fast, require very little labor, and are very accurate. Their accuracy, particularly in trace analysis, is partly due to the fact that continuous blanks are not measured in batches together with the samples, but they appear as an elevation of the baseline, which the integrator compensates for automatically, even if it is constantly drifting. Such methods are used for the determination of carbon, hydrogen, nitrogen, sulfur (CHN and CHNS), and oxygen.

The multielement determination methods, like CHN and CHNS,

TABLE I

Properties of Described Methods

Element and method		Weight of sample (mg)	Method of decomposition	Method of separation	Method of measurement	Limit of detection (μg)	Optimal amount of element (μg)
C	multielement	0.050–12	Dry dynamic-tube combustion	Gas chromatography	Thermal conductivity	0.1	400
H						0.2	50
N						0.1	100
N	micro, trace	0.5–50	Dry dynamic-tube combustion	Gas chromatography	Thermal conductivity	1	500
O		0.05–10	Dynamic-tube pyrolysis	Gas chromatography	Thermal conductivity	0.5	200
S	ultramicro	0.010–3	Dry dynamic-tube combustion + hydrogenation	Wet absorption	Spectrophotometry	0.02	10

S	trace	0.010–200	Dry dynamic-tube combustion + hydrogenation	Wet absorption	Spectrophotometry	0.1	10
S	with CHN	0.050–1	Dry dynamic-tube combustion	Gas chromatography	Thermal conductivity	2	100
					Flame photometry	0.3	5
Cl	ultramicro	0.050–0.700	Hot-flask combustion	Diffusion	Dead-stop titration	1	50
Br						2	100
I						2	200
Cl	trace	1–50	Flask combustion		Dead-stop titration	3	50
Br						5	100
I						5	200
F	ultramicro	0.050–5	Dry dynamic-tube combustion + hydrogenation	Wet absorption	Spectrophotometry	1	>5
F	trace	0.050–1000	Dry dynamic-tube combustion + hydrogenation	Wet absorption	Spectrophotometry	1	>5
P	ultramicro	0.010–40	Wet combustion	Extraction	Spectrophotometry	0.05	1–40
P	trace, macro	0.010–300	Wet combustion	Extraction	Spectrophotometry	0.05	1–40

not only save time by determining several elements in one run but also have the option of using one element—usually carbon—as an internal standard and analyzing difficult substances without weighing and without drying. This is important in the analysis of many materials, e.g., for the quality control of large series of industrial raw materials or products. Also, relationships like the sulfur–carbon and nitrogen–carbon ratios, which are very important in the analysis of ion exchange resins and many similar materials, can be determined with higher accuracy with the multielement methods than with the older single-element methods. The CHNS method is the most pronounced of the described multielement methods. It has many applications.

In the CHN method a nickel combustion tube with a larger volume and a much longer lifetime than the quartz combustion tube can be used; also, a separate reduction tube is used. Very many analyses can therefore be carried out before it is necessary to change or to regenerate any fillings. The possibility of analyzing larger samples—containing up to 12 mg of organic material—with the nickel tube is advantageous when traces of nitrogen are to be determined or when it is difficult to obtain very small representative samples. It is, therefore, convenient to use the CHN method in addition to the CHNS method, which requires more frequent attention and which cannot manage such large samples.

The automatic nitrogen determination method is very fast. One analysis takes 5 minutes, and the analyst can present a result within 10 minutes after the specimen for analysis has been obtained. The modified sampler of the instrument can be loaded with 98 or more samples. The instrument works unattended, also overnight, and prints out the results. It is particulary advantageous for agricultural and food analysis, where it compares very favorably with the automated Kjeldahl methods with regard to speed, accuracy, requirement of laboratory space, and agreeable working conditions.

A great advantage of these automatic methods is the possibility of having the instruments connected to minicomputers which not only perform the integration of the signals but also calculate the final analytical results and also can perform additional complicated calculations based upon these results. This saves many hours of awkward and painstaking calculation. Unfortunately, it is not yet possible to determine all elements with such automatic methods.

The gas chromatographic method for the determination of trace sulfur does not reach the sensitivity and the precision of the spectrophotometric combustion–hydrogenation method. Determinations of small traces of sulfur are very frequently needed. The latter method is therefore used in addition to the gas chromatographic method. The spectrophotometric measurement of the formed sulfide is made with the ethylene blue method, which is considerably more sensitive than the methylene blue method and which obeys Beer's law.

The described fluorine determination method can manage most kinds of materials, from hard minerals to oils, volatile hydrocarbons, and plant tissues. It has done extensive service in fluorine studies made by the Swedish environmental protection board.

The hot-flask chlorine, bromine, and iodine method is a pronounced ultramicro method. It cannot manage large samples. It is, however, faster and more convenient than the Schöniger-flask combustion method. The small volume of the absorption and titration solutions also provides for a very sharp endpoint of the titration when very small amounts of halogen are titrated. The electrodes, which are used for the titration, are very sturdy, and the whole titration equipment requires practically no maintenance work. The same titration method is also used for the trace determination of halogen with the described Schöniger-flask method.

The described method for the determination of phosphorus is very simple, reliable, and rational, particularly when large series of samples are analyzed, because all operations, digestion, extraction, and color development are carried out in the same vessel without transfers. The fact that the ultramicro procedure can manage samples up to 40 mg and thus can serve as a micro, semimicro, and trace procedure as well, without having to change anything but the measuring cuvette of the spectrophotometer, is also a considerable advantage.

The trace phosphorus procedure, which uses the same equipment, can manage samples up to 300 mg. It can, therefore, function as a macro, micro, or ultratrace method. In both methods the spectrophotometric measurement of the phosphorus is made with the yellow molybdophosphoric acid method, which is slightly more sensitive than the most sensitive molybdenum blue methods. The color is stable for many hours.

CHAPTER 1

REAGENT SOLUTIONS, STANDARD SOLUTIONS, AND STANDARD SUBSTANCES

Reagent solutions are used to obtain the necessary conditions for a chemical reaction to take place. They must not contain impurities that can interfere with the analysis. Contamination can arise from the atmosphere, from the walls of the container, and from tools like the pipettes used to withdraw the solution. The influence of such contamination is kept low by keeping the solutions in large, well-filled, and well-closed containers of carefully selected material and by discarding the remaining solution long before all is consumed.

Most reagent solutions can be kept conveniently in dispenser bottles, from which they can be dispensed directly to the reaction vessels. Dispensers with varying degrees of accuracy and inertness

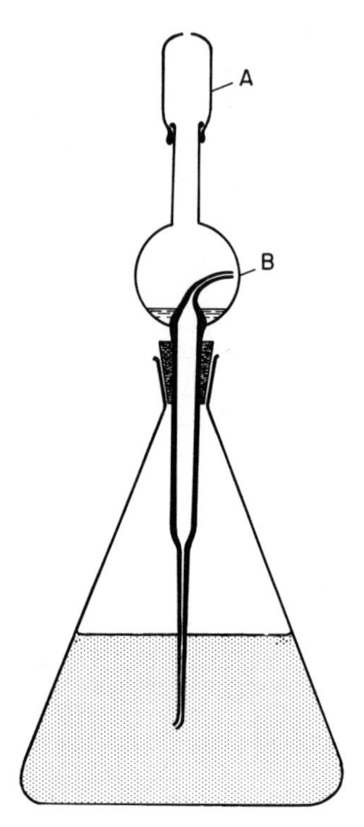

FIGURE 1. Grunbaum pipette with reagent flask. The pipette is held by a stopper of silicone rubber or some other inert material. The tip is immersed in the reagent solution. When used, the pipette with stopper is loosened, the rubber bulb A is compressed, and the hole in it is closed with a finger. Liquid is then sucked up. It washes the inner walls of the pipette and flows over into the glass bulb B. The fingertip is then removed from the hole in the rubber bulb, and the pipette is held in an inclined position. The liquid is held in the pipette by capillary attraction. The pipette is taken out from the flask, and the liquid is dispensed by closing the hole in bulb A and pressing the bulb. The liquid in bulb B, which has washed the pipette, remains there and absorbs any impurities that might diffuse in from bulb A. The tip of the pipette is always washed in the reagent solution.

and with different kinds of protection for the solutions are commercially available. The Grunbaum pipette (Grunbaum and Kirk, 1955; Grunbaum, 1970) with a flask (Figure 1) is an inexpensive, accurate, and efficient dispenser for many reagents.

Standard solutions are used as measuring solutions for titrations or for the calibration of spectrophotometric or other measuring methods. When using these solutions, it is essential to have a very accurate and well-defined concentration. The accuracy of the concentration can be impaired by evaporation of the solvent and, in a few cases, such as fluoride solutions, by contamination with aluminum or other interfering materials from the walls of glass vessels. It is therefore necessary to select the material of the containers so that such contamination is avoided. As is the case for reagent solutions, the use of large, well-filled, and well-closed containers minimizes the risk of concentration errors of standard solutions.

The influence of contamination on the purity or concentration of the solution can be further diminished by preparating concentrated stock solutions and diluting them shortly before use. The diluted solution should then be used only for a short time.

Ultramicro titrations are usually carried out with very small volumes of titrant. Evaporation of even very small volumes of solvent during the transfer of the solution to the burette or from the burette itself will therefore cause considerable change in the concentration of the titrant with corresponding errors in the analytical results. Syringe-type burettes like that shown in Figure 35 (see Chapter 19) must therefore be lubricated with Vaseline to prevent creeping of solution between the barrel and the plunger, unless the plunger has a quite tight Teflon head, and the burette should be emptied and refilled a few times before the first titration.

In our laboratory all reagent and standard solutions are prepared with water redistilled from quartz vessels. In addition, all wash bottles, which are made of polyethylene, are filled with redistilled water; whenever water is used in chemical reactions or for dilutions, it is redistilled water.

Standard substances that are suitable to be used as microanalytical standards have been reported by the Sub-Committee for Micro-analytical Standards of the Microchemistry Group of the Society for Analytical Chemistry (1962) and by the Division of Analytical Chemistry, Commission on Microchemical Techniques, (1960, 1961, 1962). For the determination of carbon, hydrogen, and nitrogen we have found benzimidazole to be a very convenient test substance; for carbon, hydrogen, nitrogen, and sulfur we use phenylthiourea. When high percentages of nitrogen linked to oxygen must be determined with high accuracy, *m*-dinitrobenzene is a convenient test compound. For oxygen we use 3, 5-dinitrobenzoic acid.

In the gas chromatographic methods it is most convenient to use standard substances with high contents of the determined elements in order to keep the relative errors low. When very low contents of the elements are determined, it is suitable to run one or two blanks together with the unknowns and to subtract the blank counts from the unknown's counts. Usually it is not necessary to subtract the blank counts from the standard's counts because their influence on the calibration factor is negligible.

For the trace determination of sulfur, halogens, and phosphorus it is convenient to dissolve accurately weighed amounts of substances like dibenzyl disulfide, trifluoroacetamide, 1-chloro-2, 4-dinitrobenzene, 4-bromoaniline, 4-iodobiphenyl, and triphenylphosphine in *weighed* amounts of dinonyl phtalate and to weigh out samples of the solutions for the calibration analyses. Use only smooth boats or other containers without folds or puckers for the analyses in order to avoid creeping and losses of solution and contamination of the pan of the balance and of other tools.

CHAPTER 2

TRANSFERS

Transfers of sample *solutions* or of aliquots of sample solutions are conveniently carried out with different kinds of pipettes or syringes (Figure 2).

Pipettes are calibrated to contain or to deliver the correct volume of liquid. If a nonwetting liquid is delivered carefully with a pipette that is calibrated to contain, the correct volume can be obtained. Otherwise, the pipette can be washed out so that the correct amount of solute is obtained.

Pipettes calibrated to deliver have a somewhat larger volume in order to compensate for the layer of water that is retained on their interior when a water solution is transferred. Since the volume of this layer is influenced by the properties of the liquid that is transferred and also by the manner of its handling, high accuracy cannot easily be achieved with small-volume pipettes of this type. Syringes avoid the wetting error. They are, however, more difficult to clean.

For the transfer of solutions to cuvettes, disposable pasteur pipettes are very convenient.

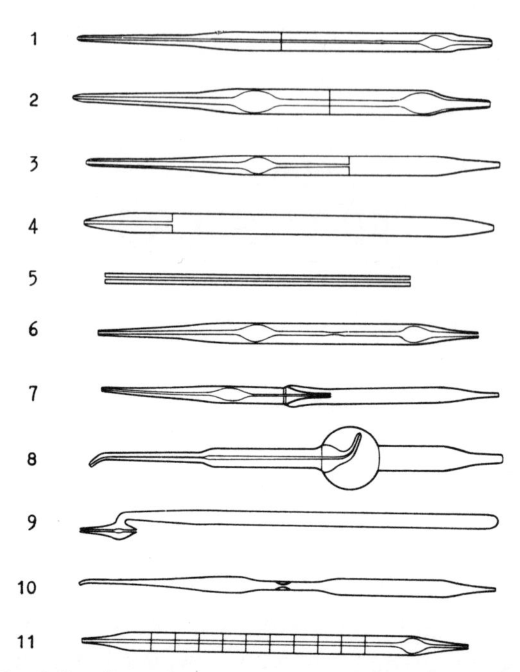

FIGURE 2. Microliter pipettes. [From Tölg (1968).] (1, 2) Pipettes, according to Sisco *et al.* (1941). (3–7, 9) Self-filling pipettes, (8) Grunbaum pipette (Grunbaum and Kirk, 1955; Grunbaum, 1970). (10, 11) Pipettes graduated to "deliver."

Solid samples are transferred to capsules and boats with the spatula shown in Figure 3a. For smeary and tough substances the solid-sample injector (Figure 3b) can be used. It is also useful for other solids, powders, etc.

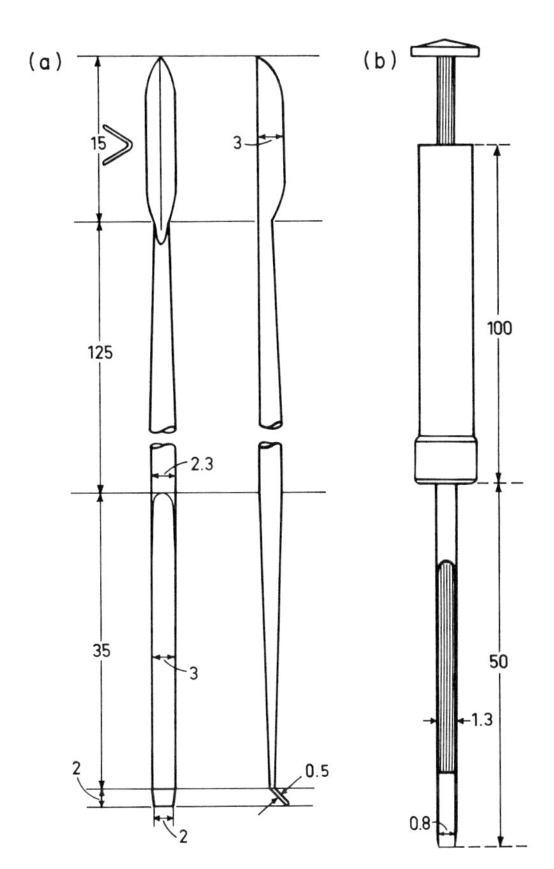

FIGURE 3. Spatula and solid-sample injector. (a) Hayman microspatula, available from Arthur H. Thomas Co. (Philadelphia, Pennsylvania). (b) Solid-sample injector, available from Scientific Glass Engineering Pty Ltd. (North Melbourne, Australia). Samples can be cut out with its tip from larger pieces, or the tip can be pressed into powders or smeary materials. The samples, collected in the tip can then be injected into capsules or boats for analysis. Dimensions are millimeters.

CHAPTER 3

HOMOGENIZATION OF SPECIMENS

Industrial and agricultural materials are very often inhomogeneous. Samples of 0.5 mg, therefore, would not be representative for an entire batch of material. Many materials, such as wheat, oats, and other grains; peas; peat; hay; wood chips; animal feeds; and dry bread, cereals, spaghetti, and other flour products can be ground in one run with an ordinary mill like the centrifugal mill ZM1 with 24 wings from Retsch (Düsseldorf, Federal Republic of Germany), using a 0.08-mm sieve.

Methods of sampling and grinding and other homogenization methods for macroanalysis—that is, analysis of samples of about 5 g—are described in the technical literature of the different branches of industry and agriculture. The following descriptions are, therefore, restricted to the homogenization of representative 5-g samples to produce representative 0.5-mg samples.

When it is important that no volatile fractions are lost from the material, a coffee mill like the KM 75 from Krups (Solingen, FRG), or the laboratory mill A10 from Janke and Kunkel (Staufen, FRG), is convenient. Most efficient in this respect is the Microdismembrator (Iyengar, 1976; Iyengar and Kasparek, 1977) from Braun (Melsungen, FRG). The Microdismembrator is a very strong, longitudinally shaking ball mill. The material to be ground is placed into an egg-shaped Teflon container together with two steel balls, which should have different diameters, usually one with a 10-mm and one with a 7-mm diameter. Many substances can be ground to a fine, homogeneous powder at room temperature. Tough, hard, smeary, or wet substances are placed in the container in the ordinary way, and it is immersed in liquid nitrogen for a few minutes. Liquid nitrogen penetrates the container, and the sample is cooled to −195°C. The container is then removed, quickly fixed in the shaker, and shaken violently. The liquid nitrogen evaporates, and even very hard, tough materials like bone, hair, or cloth are homogenized to a fine powder in 2–3 min. Rapeseeds and other oil-containing seeds, which become smeary when ground at room temperature, form a homogeneous dry powder. Meat, fresh plant tissues, and similar wet products form a homogeneous paste.

Sometimes it is important that the material to be ground does not lose volatile constituents or take up moisture. The liquid nitrogen, which penetrates the container, is quite dry, and all moisture in the sample is immediately frozen and cannot escape. When the liquid nitrogen is volatilized from the container, the sample is still so cold that no volatile constituents are volatilized. After shaking, the container is allowed to warm up to room temperature unopened so that no atmospheric moisture can condense on the cold sample. The homogenized sample has then the same moisture content and content of volatile constituents as the original material.

CHAPTER 4

DRYING OF SPECIMENS AND SAMPLES

If possible, the drying of the specimen is done before the weighing for analysis. When technical industrial and agricultural specimens are to be analyzed, drying conditions are usually specified by trade organizations or by governmental agencies. Samples of the dried specimens are then weighed out for analysis as described in Chapter 5.

Hygroscopic specimens are dried with the equipment shown in Figures 4 and 5. The drying pistol is designed so that only dry air passes into it when the stopcock is opened after drying.

Individual samples can be dried and weighed in the quartz vessels shown in Figure 4. It is usually preferable to encapsulate and weigh them as described in Chapter 5 and to dry and reweigh the crimped capsules. The crimped capsules can be kept in a tray (Figure 4) or in boats or specimen tubes in the drying pistol. Since the crimped capsules are not hermetically sealed, they must be protected from moisture and weighed quickly if they contain hygroscopic samples.

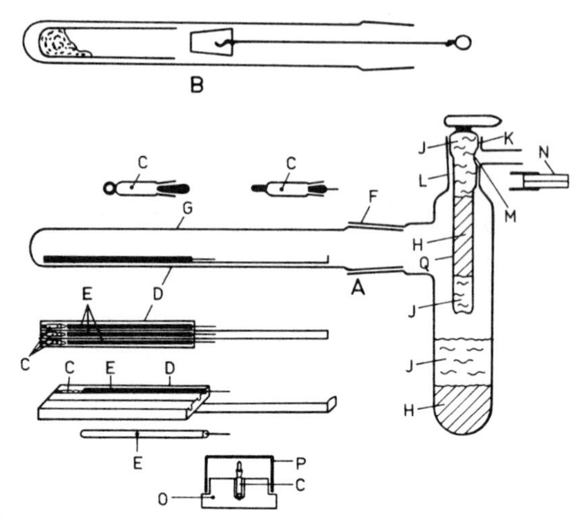

FIGURE 4. Drying pistol with accessories.]From Kirsten (1961).] (A) Drying pistol loaded with samples in quartz tubes. (B) Specimen tube with large specimen in tube of drying pistol. A long, thin corkscrew is screwed into a silicone rubber stopper and placed in front of the specimen tube during the drying. After drying and opening the pistol, the stopper is quickly inserted into the specimen tube. (C) Small quartz vessel with ground joints for drying of small weighed or unweighed samples. The vessel with the sample is placed into a booth of the tray (D) with the stopper loosely inserted and the heavy brass rod (E) behind it. The tray is then placed into the tube of the pistol, which is evacuated through the opening (M) the stopcock (K). After drying and cooling, air is allowed to pass slowly into the pistol through the capillary (N), which is attached to the stopcock (K). The pistol is then slightly inclined with its tube downward and is tapped lightly. Rods (E) slide forward and push the stoppers into the drying vessels. (L) Groove in stopcock (K), (H) drying agent, (J) quartz wool, (O) brass block, (Q) drying tube sealed to stopcock (K), (P) glass cover.

The quartz vessels are not hermetically sealed. About 0.05 μg of moisture per hour per percent of relative humidity at 25°C diffuses into them (Kirsten, 1966). They must therefore be brought into a dry atmosphere immediately after the drying. This is usually not necessary when the samples are to be analyzed or weighed immediately. The small amounts of absorbed moisture are then usually negligible.

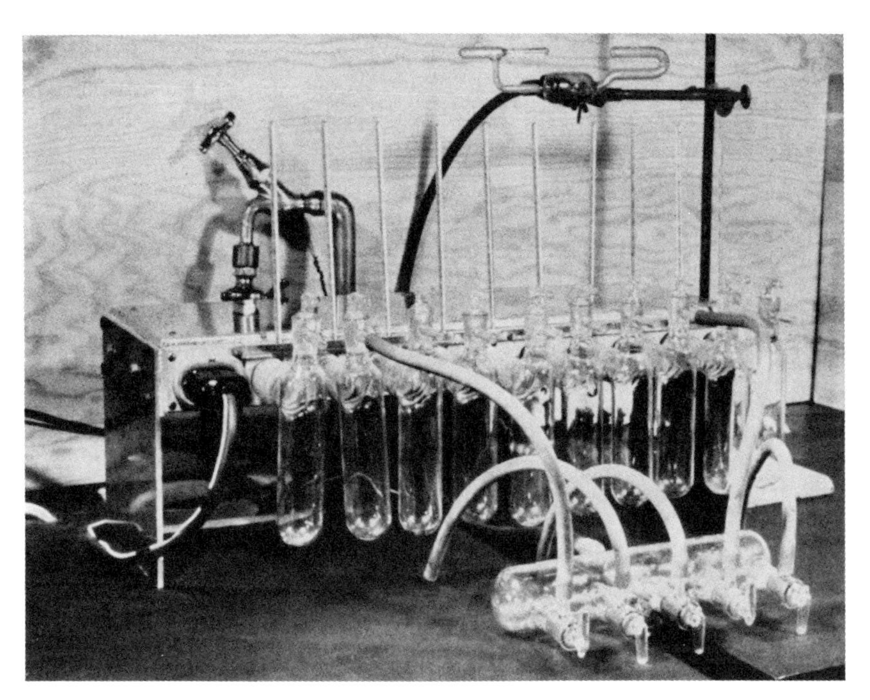

FIGURE 5. Drying pistols in gradient heating block. [From Kirsten (1961).] The block is heated with an electrical element at its left side. At the right end it can be cooled with water. The surface of contact between the water and the block is small, which makes the cooling rate almost independent of the flow rate of the water. With water-cooling, a temperature gradient of 150°C to 35°C is obtained. The samples can therefore be dried at any desired temperature. Without water-cooling the difference is only about 30°C. It is quite convenient to use two blocks without cooling at the two most common drying ranges. The drying pistols can be evacuated and kept under vacuum through a vacuum manifold with a measuring gauge, as shown in the figure.

CHAPTER 5

WEIGHING AND ENCAPSULATION OF SAMPLES*

Electronic ultramicro balances are most suitable for weighing samples. When the Cahn 27 balance is used with small pans, it is insensitive to changes in temperature, to vibrations, and even to drafts, and it can usually be placed right in the laboratory without special protection. Electrical grounding of the balance is recommended; it is also a good practice to have a grounded metal plate in front of the balance on which to place boats, capsules, and all weighing tools and to do the filling and closing of the capsules there. Before starting these operations the analyst should place his fingers on the plate in order to get rid of static charges in his own body.

Small, thin-walled metal containers should be used for the

*According to Kirsten and Kirsten (1979).

weighings because they have low buoyancy, have low moisture adsorption, and do not tend to develop static charges.

In many instances, such as determinations of moisture or ashes, the sample must be weighed and reweighed in the same container after some time and treatment. In this case the following procedure is used: Zero the balance. Weigh the container and note its weight. Zero with the automatic tare. Add the sample and read its weight. When reweighing: Zero the balance and weigh the container with the treated sample or its residue. Subtract the weight of the empty container.

When samples are weighed out for other types of analyses, it is usually not necessary to zero the balance because changes of the zero point between the taring and the weighing are negligible.

WEIGHING IN CRIMPED CAPSULES

Use the apparatus shown in Figure 6.

Capsules

Wash capsules of silver, tin, or aluminum in acetone, water, and alcohol and dry them. For trace determination of oxygen, use silver capsules and after washing and drying under hydrogen heat them to 500°C for 10 min and let them cool under hydrogen. They should not be tightly packed during the heating, otherwise they will stick together.

Procedure

Place the empty capsule on the pan of the balance and tare it. Place it on a grounded plate in front of the balance and add the

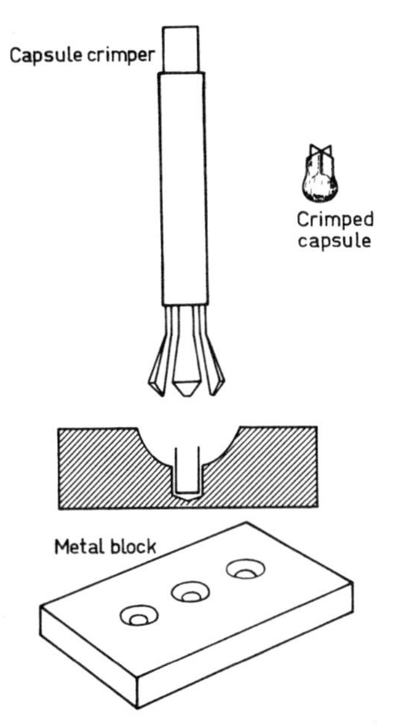

FIGURE 6. Encapsulation equipment for stable solids. [From Kirsten and Kirsten (1979).] The weighed capsule is placed in the hole of the metal block. The sample is introduced into it, and the crimper is pushed down over the capsule and crimps it. Different kinds of capsules are available from Lüdi and Cie., Metallwarenfabriken (Flawil Switzerland). Crimping equipment is available from Carlo Erba Strumentazione (Rodano-Milano, Italy).

sample. Weigh the capsule and crimp it. When the samples are to be used in the sampler of the Carlo Erba analyzer, fold down the "wings" of the crimped capsule in the cavity of the metal block to prevent it from jamming in the sampler.

HYGROSCOPIC SAMPLES

Use the apparatus shown in Figures 6, 7, and 8.

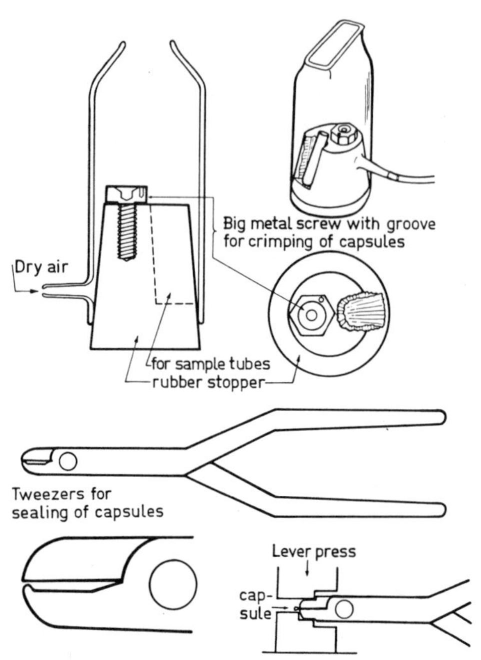

FIGURE 7. Arrangement for weighing hygroscopic and volatile samples. [From Kirsten and Kirsten (1979).] Glass bell mounted on rubber stopper. Inner diameter of bell 45 mm, height 120 mm. Height of stopper 35 mm. The wide hole in the screwhead is used for different kinds of folded or smooth-wall capsules. The narrow hole, diameter 2.5 mm, depth 3.5 mm, is for narrow capsules for liquids, according to Pella and Colombo (1973). We have several stoppers with screws with different sizes of holes that fit into the same glass bell. The jaws of the tweezers must be short in relation to the handles, to provide for a strong pressure, and they must be well polished to avoid damaging the capsules. The lever press is Model APK-TBI from Berg and Schmidt (Stuttgart, Federal Republic of Germany). The jaws of the tweezers in the press are flat and well polished. The capsules are Lüdi 84 0176 01, tin, inner diameter 1 mm, length 5.5 mm, and wall thickness 0.07 mm, designed by Pella (1978).

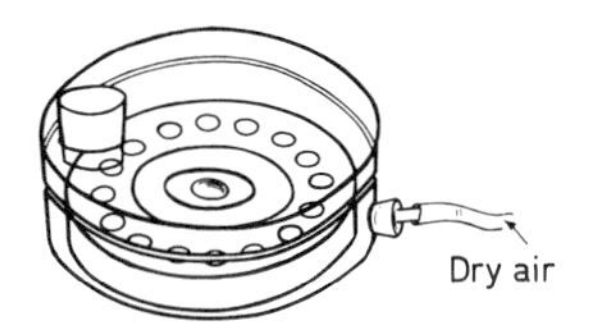

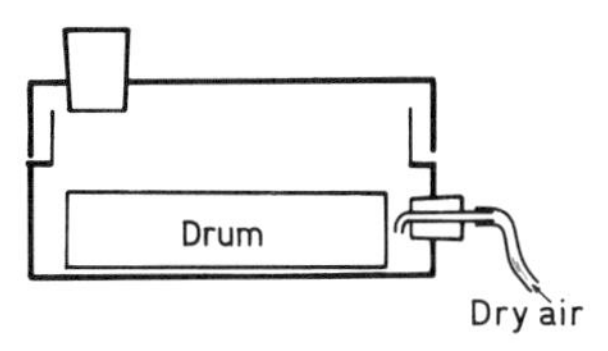

FIGURE 8. Drum jar of glass for encapsulated substances. [From Kirsten and Kirsten (1979).] The lid is ground on the jar, but is not lubricated. The silicone rubber stopper is placed loosely in the hole in the lid. It is taken away and the capsules are introduced through the hole into the drum, against the flow of dry air.

Procedure

Place the empty capsule on the balance and tare it. Place it in the cavity of the screwhead in the glass bell and place the sample tube in the groove of the stopper in the bell and open it. Transfer the sample to the capsule and crimp the capsule. Weigh the capsule quickly. If it must be kept free from moisture after the weighing, put it into the drum jar shown in Figure 8 or into another container under dry gas before the analysis.

The crimped capsules are not hermetically sealed. Some moisture will diffuse into them during and after the weighing. The rate of diffusion can be diminished by further compressing the wings of the crimped capsules with a pair of tweezers.

The moisture uptake of capsules closed with different devices is shown in Figure 9.

The weighing of a capsule on a Cahn 27 electronic balance takes 30–50 s. This means that the weight increase caused by diffused moisture can generally be neglected.

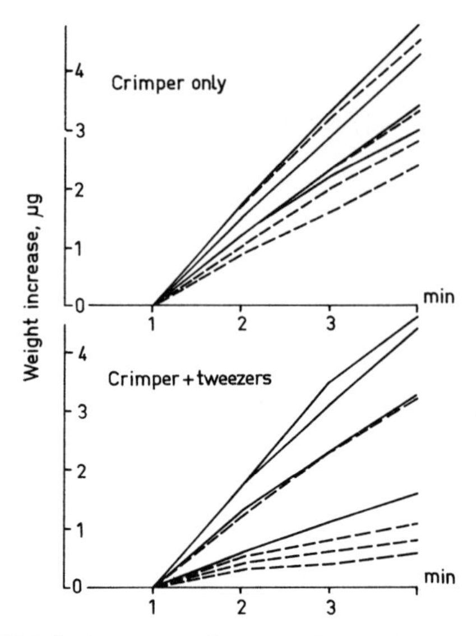

FIGURE 9. Weight increase of crimped capsules containing dry silica gel at 22.5°C and relative humidity of 19.5%; ——Sn capsules, --Ag capsules. [From Kirsten and Kirsten (1979).]

If it is assumed that the relative humidity inside the capsule is zero and that the amount of diffusing moisture is proportional to the absolute humidity of the atmosphere, an approximate correction for this moisture can be made in cases where very high accuracy is necessary.

If the sample does not tolerate any moisture at all, it can be encapsulated and weighed like a volatile solid, as described in the following.

VOLATILE AND HYGROSCOPIC LIQUIDS

Use the apparatus shown in Figure 7. The encapsulation must be made airtight. Since a part of the gas in the capsules is also enclosed, the encapsulation must be made under nitrogen when oxygen is to be determined and under oxygen when nitrogen is to be determined. Substances sensitive to oxygen can be encapsulated under helium.

Procedure

Place the narrow, thick-walled capsule in the narrow hole in the screwhead and introduce the liquid sample with a syringe. Seal the opening with tweezers.

Sometimes solids can also be encapsulated in this manner. In this case the samples are introduced into the capsules with a solid-sample injector (Figure 3).

With the method described some evaporation can occur in highly volatile samples before the capsule is sealed. If the sample is a mixture of compounds having different boiling points, this can cause some fractionation and erroneous results. This can be avoided in the following manner: Use very narrow capsules (e.g., Lüdi 84 0176 01), fill them to overflowing with a syringe, and seal them with flat tweezers near the bottom as shown in Figure 7. Any liquid from which evaporation has occurred is squeezed out, and only the innermost liquid, which has its original composition, remains in the capsule.

In order to be sure to obtain an airtight seal it is recommended that the pressure of the tweezers be increased by compressing their jaws between the jaws of a hand-lever press, as shown in Figure 7. When very difficult materials, like oil-containing mixtures, are encapsulated, the capsule can be sealed with the spot welder (Figure 10). No protective gas is needed with this method.

VOLATILE SOLIDS, SMEARY AND OILY MATERIALS, AND WET SOLIDS

Use the equipment shown in Figure 10 together with that in Figure 7.

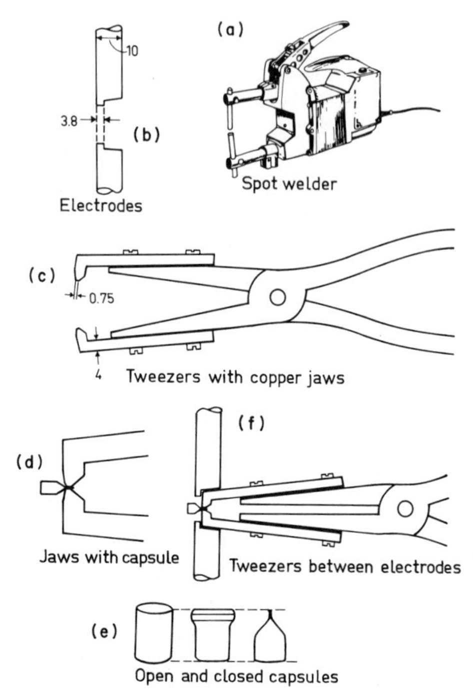

FIGURE 10. Spot welding equipment for encapsulation of volatile solids, smeary and oily materials, and wet solids. [From Kirsten and Kirsten (1979).] (a) Electric spot welder, TECNA 7930, 2 kV A, with regulators for time and power, from TECNA, Castel S. Pietro Terme (Bologna, Italy). (b) Electrodes of spot welder, shaped to fit to jaws of tweezer (c). (d) Jaws of copper, 9 mm broad, screwed to the tweezers with insulating layers in between. (e) Tin capsules, height 6 mm, diameter 2.9 mm, wall thickness 0.06 mm, Lüdi 84 0180 41. (f) Tweezers with capsule between electrodes. The electrode arms of the spot welder are cut off to a total length of 7 cm in order to increase the pressure exerted on the tweezers. The teeth of the tweezers must be well rounded and polished to prevent tin from the capsules from sticking to them. The polishing is conveniently done by rubbing the teeth with a safety pin. Dimensions are millimeters.

Procedure

Weigh the tin capsule and introduce the sample into it with the equipment shown in Figure 7 under protecting gas. Grasp the capsule near its opening with the copper tweezers (Figure 10). Introduce the tweezers between the electrodes of the spot welder and press down the lever of the welder. The sample is now hermetically sealed into the capsule and can be handled without a protecting gas.

PRECAUTIONS

When partially volatile samples are hermetically encapsulated for analysis, it is advisable to wash the capsules with acetone or another suitable solvent after the sealing in order to remove any residues that might adhere to the walls above the seal.

CHAPTER 6

SPECTROPHOTOMETRY

In ultramicro analysis and trace analysis it is frequently necessary to measure very dilute solutions. Sometimes it is possible to concentrate the analyte through extraction, but this requires an extra operation and increases the risk of contamination. It is, therefore, desirable to have a spectrophotometer that can accommodate long cuvettes.

To get good results with long cuvettes the light beam must be masked so that it does not strike the cuvette walls. Since reflection inside the cuvettes is unavoidable, the geometry of the sample cuvette must be the same as that of the reference cuvette. The holes for filling the cuvette must not be blown but drilled. Blowing will not give a reproducible geometry. To avoid stray light the walls of the cuvettes should be made of black glass or quartz.

When measurements are made with long cuvettes, it is important not to touch the walls of the cuvettes with the fingers. The parts of the cuvettes that are touched will become warm, which will cause schlieren in the solution. This can completely spoil the measurements. The risk is greater with organic solvents than with water solutions.

It is also possible to use fix flow-through cuvettes. Since all adjustments and measurements are made under the same conditions in this case, errors will largely be compensated for.

When a color is extracted from a water solution into a water-immiscible solvent, the solvent is saturated with water. A very slight cooling of the cuvette will then cause formation of hazes of water, which will spoil the measurement, particularly in long cuvettes. This risk is especially high with very water-immiscible solvents such as carbon tetrachloride or benzene. After shaking, the system, therefore should be cooled first, shaken again for a few seconds, and then centrifuged quickly to clear the organic phase.

CHAPTER 7

MULTIELEMENT DETERMINATIONS

Multielement determination methods are advantageous in several respects: Since several elements are determined in the same run of the same sample, time, work, and materials are saved. Another advantage is the possibility of using one of the elements as an internal standard. This means that the amounts of the determined elements are related not to the weight of the sample but to an element with a negligible variation. For example, the carbon content of most dry cereals is quite constant. Protein determinations can therefore be made by analyzing an undried, unweighed sample and relating the nitrogen content to the carbon content. The moisture content of the sample can be calculated simultaneously from the hydrogen figure. Many raw materials and industrial, technical, and agricultural products (e.g., organic chemicals, plastics, rubber products, coal, peat, grain, mineral and fat oils, sugar and flour products, milk, cheese, solvents, and pharmaceutical and technical preparations) can thus be analyzed quickly without weighing. The nitrogen, sulfur, or hydrogen contents are related to the carbon figures. In the case of dilute water solutions the carbon, nitrogen, and sulfur figures are related to the hydrogen figures. Hygroscopic and volatile materials can be analyzed quickly and accurately without drying and without painstaking weighing procedures.

Methods and instruments for the calculation of atomic relationships and formulas of compounds analyzed without weighing with the Carlo Erba CHN or CHNS analyzer have been described by Haberli (1973) and Colombo *et al.* (1979).

Metals, halogens, sulfur, and phosphorus can be brought into an ionic form with a flask combustion method, and then several of them can be determined simultaneously through ion chromatography (Small *et al.*, 1975; Smith *et al.*, 1977).

CHAPTER 8

THE CARLO ERBA AUTOMATIC GAS CHROMATOGRAPHIC ELEMENTAL ANALYZER*

The Carlo Erba automatic gas chromatographic elemental analyzer uses flash combustion in oxygen for the determination of carbon, hydrogen, nitrogen, and sulfur. The gaseous combustion products are separated by gas chromatography and measured with a hot-wire detector and an integrator. In the oxygen determination the sample is pyrolyzed over nickelized carbon, and the gaseous reaction products are also separated by gas chromatography and measured in the same manner.

That such analyses can be carried out with very small samples with a higher accuracy than that obtained with the conventional

*See Poy (1970).

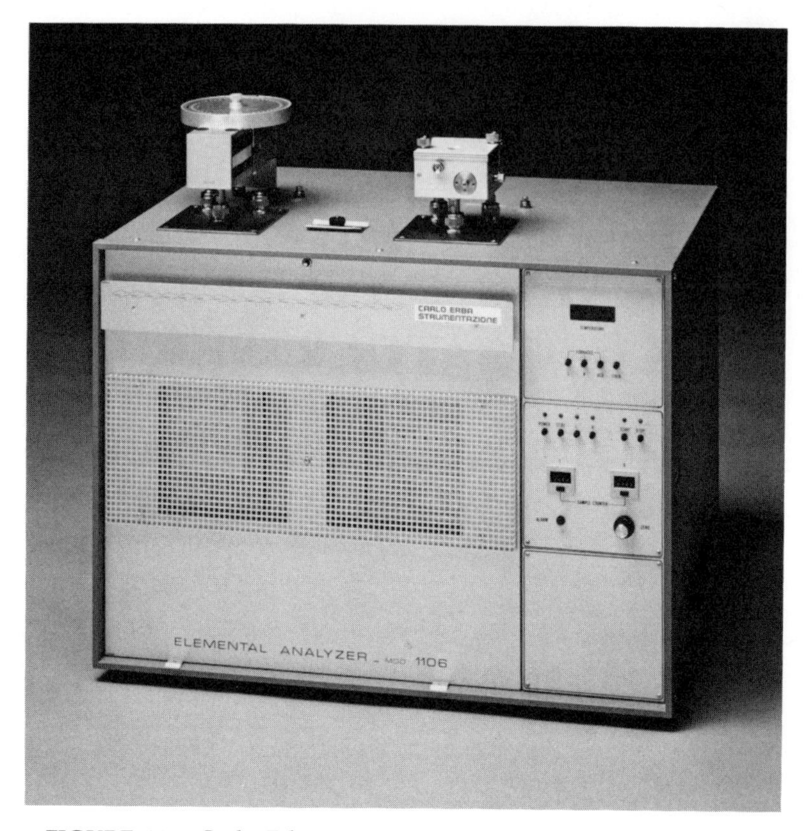

FIGURE 11. Carlo Erba 1106 automatic elemental analyzer with 50-position sampler, left, and 23-position sampler, right.

microanalytical methods is mainly due to the great improvements in chromatographic instrument design, such as accurate gasflow control and linear detectors, made in recent years.

It is a great advantage of the chromatographic separation method over the conventional chemical separation methods that volatile interfering compounds, which can be carried through the chemical scrubbers, are efficiently filtered away by the chromatographic column. Also, when chemical separation methods are used, the fillings of the scrubbers must be renewed frequently. A Porapak QS column for CHN determination can be used for several years of continuous work.

The analyzer, model 1106, is shown in Figure 11, and its flow scheme is shown in Figure 12. It has two channels for combustion analysis, in which oxygen is added to the carrier gas. In the left channel the addition of oxygen can be shut off so that this channel can also be used for the determination of oxygen by pyrolysis.

There is also an older model, 1104, of the analyzer, which has no arrangements for the addition of oxygen in the left channel. This channel can, therefore, only be used for the determination of oxygen. The right channel has the same functions as that in model 1106.

There are two types of samplers: the first, with 50 holes, fits only

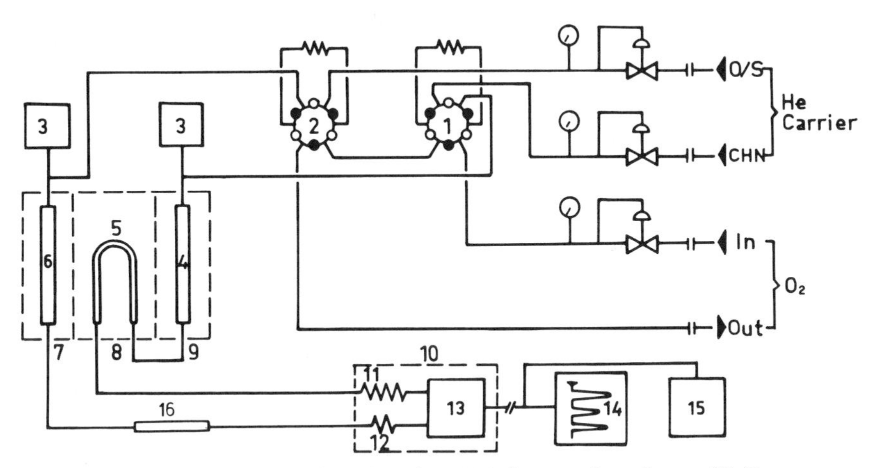

FIGURE 12. Layout of Carlo Erba 1106 elemental analyzer. (1) Oxygen injection valve, CHN. (2) Oxygen injection valve, S or CHNS. (3) Samplers, (4) Combustion tube, CHN. (5) Reduction tube, CHN. (6) Pyrolysis tube, O; or reactor tube, S or CHNS. (7) Reactor furnace, O; or S or CHNS. (8) Reduction furnace, CHN. (9) Combusion furnace, CHN. (10) Thermostatic Oven. (11) Chromatographic column, CHN. (12) Chromatographic column, O, resp. S or CHNS. (13) Thermoconductivity detector. (14) Recorder. (15) Integrator. (16) Scrubber.

model 1106, whereas the other, with 23 holes, can be used for both models.

With the 23-hole sampler the drum is filled with samples and introduced into the sampler, where it is under helium gas and unattainable during the analytical run. In the 50-hole sampler the drum remains in the laboratory atmosphere and samples can be added or removed during the run. Hygroscopic substances must be sealed hermetically to prevent uptake of moisture.

The analyzers are provided with peak sensors, which turn off the instrument when no peak, or only a very small peak is eluted in an analysis, assuming that either something is wrong or the series of analyses is finished. This makes it impossible to carry out blank and trace determinations. Since such determinations frequently are necessary it is recommended the peak sensor be inactivated and a timer to turn off the instrument be used.

THE FUNCTION OF THE CHN CHANNEL

For the CHN determination the automatic analytical cycle begins with the injection of a few milliliters of oxygen into the helium carrier gas stream by means of the oxygen injection valve (1) (Figure 12). After a few seconds the sampler (3) drops a sample into the combustion tube (4), where it is burned. The gas mixture—which contains carbon dioxide, water, nitrogen, nitrogen oxides, and eventually combustion products of other elements from the sample—passes through a layer of chromium oxide and cobaltic oxide–silver, where complete oxidation is achieved and halogens and sulfur are retained. It then passes into the reduction tube (5), which contains metallic copper at 650°C. The copper retains the excess of oxygen and reduces nitrogen oxides to elemental nitrogen.

The gas mixture then passes through the chromatographic column (11), where its individual components are separated, into the hot-wire detector (13). The signals from the detector go to the recorder (14) and to the integrator (15), which prints out the peak areas for every separated gas. The apparatus is calibrated with microanalytical standard substances.

THE FUNCTION OF THE OXYGEN CHANNEL

The samples are in the left automatic sampler (3) (Figure 12), which, with preset intervals, drops them into the pyrolysis tube (6), which contains a contact filling of nickelized carbon at 1020°C–1060°C. The sample is pyrolyzed instantaneously, and the carrier gas, helium loaded with chloropentane vapor, sweeps the oxygen-containing products of the pyrolysis through the carbon contact filling, with which all the oxygen reacts to form carbon monoxide. The gas mixture—containing hydrogen, nitrogen, carbon monoxide, volatile nickel compounds, and possibly other interfering compounds—passes through the scrubber (16), where interfering substances are retained, through the chromatographic column (12), where its individual components are separated, into the hot-wire detector (13). The signals from the detector go to the recorder (14) and to the integrator (15), which prints out the peak area of the carbon monoxide peak. When a column with high separation power is used, the hydrogen and nitrogen peaks can also be measured. These peaks can, however, give only a rough indication of the hydrogen and nitrogen contents of the sample. The apparatus is calibrated with microanalytical standard substances.

PROCESSING OF THE DATA

The peak areas are measured with normal integrators for chromatography. The instrument is calibrated with pure standard compounds. In order to minimize the influence of small accidental measuring errors on the calibration factors, the calibration compounds should have high contents of the measured elements. When traces are to be determined, it is advisable to carry out blank runs between the unknowns and to subtract the blank areas from those of the samples. Blanks can be empty runs or, better, runs of substances with a composition similar to that of the unknowns except for the absence of the trace element to be determined. The influence of the blank on the calibration factor is usually negligible.

When high analytical accuracy is required, calibration substances should be run in intervals between the unknowns—say, after every tenth unknown. The averages of the calibration factors—gross errors being rejected—are then used for the calculation. If the factors drift in the course of a series of analyses, individual factors for the unknowns can be interpolated. This hardly ever happens with the described instruments.

Computerization of the calculation is very desirable. Different procedures have been described or proposed (Stoffel, 1972; Haberli, 1973; Howarth, 1977; Colombo *et al.*, 1979). Carlo Erba Strumentazione sells a microprocessor, model 2000 C, which we have used with the Elemental Analyzer Model 1106 for a few years and which saves us much calculation. Haraldsson (1980) uses a commercial low-priced computer, which can be attached to any model.

The ideal data processor should have the following features: It should accept and process signals from all models of analyzers, and it should be able to accept signals from more than one detector at the same time. It should be able to accept the signals from the electronic balances, and it should be able to carry out all the calculations mentioned above. Also, when multielement analyses are carried out without weighing, it should be able to base the calculations on relationships between the elements rather than use the weight of the sample. No data processor with all these abilities is yet available.

SOME COMMON OPERATIONS

The encapsulation and weighing of samples has been described in Chapter 5. It must also be observed, however, that in order to avoid jamming of the capsules in the sampler, the wings of crimped capsules must be pressed to its main body so that they cannot creep in between sliding parts of the sampler. This is done with forceps after the crimping.

Oblong capsules must be either so long that they cannot tip over and jam or so short that they will not jam even when lying.

When changing a gas tank, always turn off the oxygen supply and the detector current first and wait until the oxygen pressure has gone down. Columns and detectors must never be left in contact with oxygen or air when hot.

For short-time shutdown of the instrument push the standby button, turn off the detector current and the oxygen flow, and decrease the carrier gas flow to a very small rate—about 5 mL/min. Check that there is enough gas in the tanks to keep up the pressure during the shutdown period.

For long-time shutdown turn off the detector current and the main line and the oxygen supply. When the instrument is cold, close all gas outlets and turn off the helium and the service gas supplies and the cooling water.

GENERAL REMARKS

The Carlo Erba elemental analyzers give excellent performance when properly treated. Some points must be observed: *Detectors and columns* must be protected from contact with oxygen or air when hot. *Make sure therefore that the service gas pressure never goes* *below 3 kp/cm²* while the oxygen flow is on.* This would cause the oxygen to pass freely through the instrument, which would damage reduction tubes, separation columns, and detectors of both channels and would burn away the carbon of the oxygen reactor tube.

In order to keep servicing times as short as possible and decrease the risks of getting air into the columns, it is most convenient to have reduced reduction tubes, prefilled scrubbers, and combustion tubes without tin oxide residues ready. It then takes only a short time to remove the old tubes and scrubbers and replace them with the new ones. The consumed reduction tubes can then be reduced, the scrubbers can be refilled, and the tin residues can be removed from the combustion tubes when it is convenient.

The instruments can be supplied with two samplers, one with 23 positions and one with 50 positions. The newest 50-position sampler with two-step action is very efficient. In its body the encapsulated sample is transported with a plastic slide. If it happens that the capsule is jammed inside the sampler, this slide can be scratched and the sampler can then lose its tightness. The slide must then be replaced with a new one. It is therefore important to shape the capsules so that no jamming can occur. The 23-position sampler is very sturdy and gives excellent service without special precautions.

*1 kp/cm^2 (kilopond/square centimeter) = 14.223 psi.

CHAPTER 9

AUTOMATIC SIMULTANEOUS DETERMINATION OF CARBON, HYDROGEN, AND NITROGEN*

APPARATUS

Use the CHN channel of the Carlo Erba analyzer (Figures 11 and 12) with the reactor tubes shown in Figure 13.

REAGENTS

Combustion Catalysts

Granulated chromium oxide is available from Carlo Erba Milano, Italy, or can be prepared in the following way: Dissolve 60 g of

*See Poy (1970); Pella and Colombo (1973).

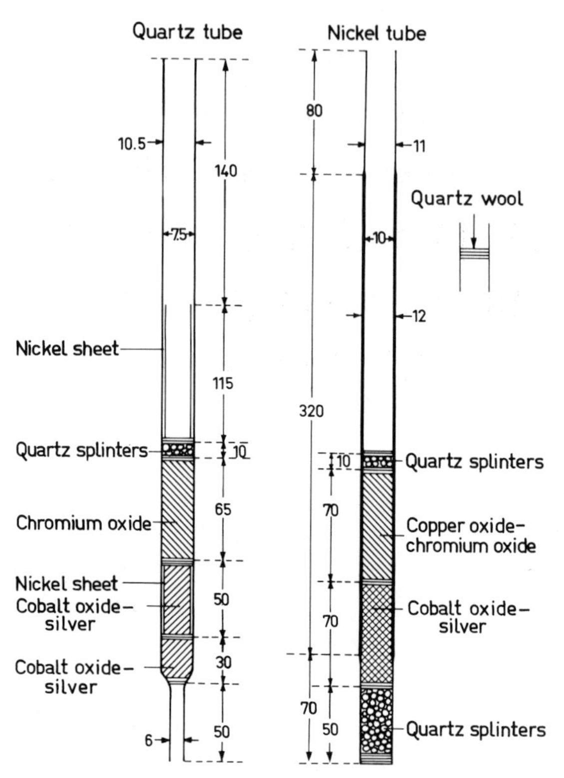

FIGURE 13. Quartz tube and nickel tube for CHN determination. The rolls of 0.1-mm nickel sheet are made to overlap about 5 mm. They are rolled on a metal rod slightly wider than the inner diameter of the quartz tube and pushed into place with a sharp-edged 7-mm metal rod, keeping the rod touching the wall of the quartz tube at the place where the nickel sheet overlaps. No cobalt oxide–silver should get outside the nickel sheet.

The nickel tube has a considerably higher oxidation capacity and a much longer lifetime than the quartz tube, and it can receive many more samples before the residues must be removed. Dimensions are millimeters.

chromium(VI) oxide, puriss. in 50 mL of water. Add 400 g of chromium(III) oxide powder and mix to get a paste. Transfer it to a nickel plate and dry at 180°C. Break it into lumps before it is quite dry. When dry, crush and sieve it to a particle size between 0.5 and 1.5 mm. Add the fines to the next batch. Heat the granules in a quartz tube to 900°C in a flow of air or oxygen.

Cobalt oxide–silver, *ca.* 10% Ag, can be prepared in the following way: Dissolve 85 g of silver nitrate, $AgNO_3$, puriss., in 100 mL of water and add 0.5 kg of cobalt oxide, Co_3O_4 powder, pro analysi. Mix well and dry at 150°C in a nickel dish, stirring now and then to get lumps of convenient size. Break up and crush the lumps and sieve to a particle size between 0.5 and 1.5 mm. Add the fines to the next batch. Heat the grains in an alumina tube under oxygen to 1100°C for at least 2 hours and let them cool very slowly under oxygen.

To get metallic copper from wire use Merck 2701 (Darmstadt, FRG), or reduce copper oxide, Merck 2765, with very pure hydrogen. Use quartz wool, quartz splinters 1.0–1.5 mm, nickel plate 0.1 mm.

Alternative Combustion Catalysts

To obtain chromium oxide–copper oxide, *ca.* 24% Cu_2O, place 100 g of granulated chromium oxide, Cr_2O_3, in a procelain dish and heat it to 150°C. Dissolve 110 g of copper sulfate pentahydrate, $CuSO_4 \cdot 5H_2O$, in 100 mL of water, with heating to near boiling. Wet the hot chromium oxide with a part of the hot copper sulfate solution. All the solution should be taken up by the grains. Dry at 150°C, stirring now and then; wet again with more solution and dry. Repeat the process until all copper sulfate is absorbed in and on the granules of the chromium oxide. Place the dry granules into an alumina tube and heat them in a flow of air or oxygen to about 1000°C until no more sulfur trioxide is liberated. Sieve off all dust.

To obtain granulated nickel oxide, mix 1 kg of green precipitated nickel(II) oxide with 150 g of nickel(II) sulfate·$7H_2O$. Add water slowly while stirring until a tough paste is obtained. Dry at 150°C and crush and sieve the resulting cake. Heat the grains between 2.0 and 0.5 mm in a slow flow of oxygen to at least 1100°C for at least 2 hours. Add the fines to the next batch.

Granulated cobalt oxide is available from Carlo Erba.

ADJUSTMENT OF APPARATUS

Assemble the apparatus according to the manual. Fill the combustion tubes according to Figure 13. Fill reduction tubes according to the manual, but only with metallic copper, not with copper oxide, or make reduction tubes of quartz and fill them in the same manner.

Be careful to fill the combustion tube and the reduction tube only with rather coarse material, avoiding fine dust. Do not compress the quartz wool too tightly, and check that there are no restrictions anywhere that prevent a reasonably free gas flow between the membrane valve of the helium carrier gas and the chromatographic column. Restrictions can cause small peaks when the excess of oxygen is absorbed in the reduction tube at the beginning of the chromatogram, which can interfere with the zeroing of the integrator.*

It is very convenient to introduce a small restriction capillary between the separation column and the detector and to adjust it so that the gas flow rate of 30 mL/min is obtained at an overpressure of about 1 kp/cm^2. This is easily achieved by introducing a 0.2-mm metal wire into a 0.25-mm stainless steel capillary, available from

*The newest Carlo Erba instruments are provided with flow regulators. They must not be closed too tightly. If a peak appears before the nitrogen peak, open the flow regulating valve wider and regulate the gas flow with the pressure regulating valve.

Carlo Erba, and bending its end into a hook to prevent it from sliding completely into the capillary. Adjust the length of the capillary and wire until the desired flow rate is obtained at room temperature with a pressure of about 0.7 kp/cm^2.*

This arrangement has the advantage of allowing the gas flow rate to be determined mainly by the gas pressure and the capillary, and the same gas pressure and the same inlet delay time can always be used with the instrument, independently of changes of combustion tubes, reduction tubes, and columns.

If the nickel combustion tube is used, the 6-mm connection to the reduction tube must be replaced with a wider one. The tube is introduced into the furnace from below, and when hot, it must be handled with gloves and suitable forceps or tongs. Since the surface layers of the nickel contain small amounts of carbon, it must be kept hot for a few hours with oxygen passing through it occasionally, in order to oxidize the surface layers and remove the carbon, before connecting the combustion tube to the reduction tube.

The nickel combustion tube has a somewhat larger volume than the quartz tube. It can, therefore, receive more samples before the ashes must be removed. Its lifetime is many times that of the quartz tube. In addition to the ordinary chromium oxide filling, other catalysts, such as cobalt oxide, nickel oxide, or copper oxide on chromium oxide, can be used without protecting the walls of the tubes. The three last-named catalysts function as oxygen donors, and they increase the oxidation capacity of the tube considerably. Samples of up to 12 mg of benzimidazol containing 71.17% of carbon, 5.12% of hydrogen, and 23.71% of nitrogen can be analyzed using a 10-mL loop for the addition of the oxygen. With samples larger than 4 mg of this substance the chromatographic

*Pella (1982) reports that the correct flow rate is obtained with a capillary of 220-mm length and 0.25-mm inner diameter. The newest Carlo Erba instruments are provided with a restriction capillary after the detector. No further restriction should, therefore, be necessary.

separation between the nitrogen peak and the carbon dioxide peak is no longer complete. However, since these peaks are almost symmetrical an ordinary gas chromatographic integrator computes an accurate separation also with these large samples. Good chromatographic separation is obtained with cereals and other food products with lower nitrogen contents, even with quite large samples.

When large samples are analyzed with a 10 mL oxygen loop it is important that the time during which the carrier gas passes through the loop is sufficient to sweep out all its oxygen. When a gas flow rate of 30 mL/min and a gas pressure of 1 kp/cm^2 is used, the loop contains 20 mL of oxygen, and it takes 40 s to sweep it out. In the 1104 instrument the closing of the loop is controlled with a separate timer, which can be adjusted to the right time. In the 1106 instrument the loop is closed when the sample falls. With the ordinary arrangements this time is too short. This can be remedied by introducing a 10-mL delay loop into the carrier gas line after the bimatic valve. The inlet delay time can then be determined in the ordinary way described below, and it will be long enough to allow all oxygen to be swept out from the loop.

When analyzing large samples, it is also important that the connections between the combustion tube and the reduction tube and between the reduction tube and the separation column are kept at above 100°C in order to avoid a broadening of the water peak. When nickel oxide is used as the combustion catalyst and halogen-containing materials are analyzed the connecting tubes should be held at about 200°C. Some hygroscopic nickel halide sublimes into the tubes and retains moisture at lower temperature. If tailing of the water peak still occurs, it can be corrected by replacing the connecting tube between the combustion tube and the reduction tube with a new one. The ordinary 2-mm steel tubing is quite convenient for this purpose.

Adjust the temperature of the combustion furnace to 1030–1050°C and that of the reduction furnace to 650°C.

It is convenient to dope the carrier gas with water vapor. Fill the scrubber tube at the entry with copper sulfate pentahydrate, $CuSO_4 \cdot 5H_2O$.

The volume of the oxygen injection loop is 5 mL. Hence when the oxygen manometer shows 1 kp/cm^2, 10 mL of oxygen at atmospheric pressure are injected. One mg of organic material consumes at most 3.1 mL of oxygen, and 10 mg of tin consumes 2.1 mL. Calculate the oxygen demand of samples plus capsules and adjust the oxygen pressure so that at least 2 mL is added in excess.

Put a few empty capsules into the drum and analyze them according to the manual, using different sample inlet delay times (say, 17, 20, 25, and 30 s), and observe the flashes in the combustion tube. Adjust the inlet delay to about 3 s before the maximum flash.

There may be some organic impurity in the reduction tube (e.g., when the copper has been reduced with impure hydrogen). In this case nitrogen blanks are obtained with nitrogen-free samples, but no nitrogen blanks are obtained without samples: Carbon dioxide and water from samples react with the carbon from the impurity to form carbon monoxide and hydrogen, which are measured as nitrogen. The error can be eliminated through doping the carrier gas overnight with water by placing a wad of wet cotton into the sampler. The water vapor will eliminate the carbonaceous impurity

$$CO_2 + C = 2CO$$

and

$$C + H_2O = CO + H_2.$$

If the exit end of the reduction tube is filled with copper oxide, no nitrogen blanks will be obtained with nitrogen-free substances even if the reduction tube is contaminated, because the carbon monoxide is oxidized to carbon dioxide and the hydrogen to water. High-carbon results will then be obtained, although the instrument shows no blanks in empty runs. It is therefore preferable to fill the reduction tube with copper only.

PROCEDURE

Carry out the analyses according to the manual. Use benzimidazole or a similar compound to determine the factors for carbon, hydrogen, and nitrogen. When the content of one or more of these elements in a sample is extremely low but the element must be determined accurately, analyze, if possible, a compound with a similar composition, but without the element in question, between the analyses of the unknown. Then subtract the "blank" counts from the sample counts. The influence of the blank on the factor is usually negligible.

When large samples are analyzed using an oxygen supplying catalyst like nickel oxide at a temperature above 1000°C, the catalyst sinters slowly and looses a part of its activity. This causes formation of methane in the combustion. The methane is measured as nitrogen. It is, therefore, recommended to carry out an analysis of a nitrogen-free substance like sucrose now and then. When unacceptably high nitrogen blanks are obtained the nickel oxide must be replaced with a new charge.

When the flash compartment of the combustion tube is filled with residues from capsules and samples so that additional samples will not become hot enough to give efficient flash combustion (which can be observed through the window in the sampler), take out the tube and scratch out the residues with a steel hook until the quartz splinters begin to appear. Add new quartz splinters and quartz wool and replace the tube in the apparatus.

Oxydize exhausted reduction tubes thoroughly, first with air and then with oxygen at 800°C, and reduce them slowly with very pure hydrogen at about 300°C. The copper sinters. This does, however, not impair its efficiency, nor does it seriously impair the chromatographic separation.

When water doping is used, blank runs show a water peak followed by a tailing negative peak with the same area: The oxygen, which is injected, is dry. On its way through the system it takes up water from the walls of the tubes and vessels. When it is absorbed in the reduction tube, this water is concentrated in the carrier gas and produces a peak. The carrier gas, which follows the oxygen, loses a part of its water to the walls, which have been dried by the oxygen. This causes the negative peak.

The water doping decreases tailing and the next analysis can be started sooner.

When the water peak in the analysis covers both the positive and the negative blank peaks, these peaks compensate for each other and no error is made. When very low contents of hydrogen must be determined very accurately and the negative peak is not covered, the positive peak of the blank run must be subtracted as a blank.

When analyzing mineral oils use cholesterol or hexacosane for the calibration of the instrument.

APPLICABILITY

The method is used for the analysis of pure organic compounds as well as for industrial and agricultural raw materials and intermediate and final products such as rubbers, plastics, oils, coal, peat, shale, soils, grains, pharmaceuticals, and various food products.

Large samples of low oxygen demand for combustion (e.g., 20 mg of soil) can be analyzed.

The high temperature created by the combustion of the tin capsules ensures a complete decomposition of metal salts and other refractory materials.

Organic fluorine compounds, which contain hydrogen, give correct results. Polytetrafluoroethylene gives correct carbon results, but it also gives a peak of silicon tetrafluoride, which is measured as 3–6% nitrogen. According to Colombo and Giazzi (1982) this error does not occur if 8% of tungsten(VI) oxide on alumina is used as the combustion catalyst instead of chromium oxide.

CHAPTER 10

AUTOMATIC SIMULTANEOUS DETERMINATION OF CARBON, HYDROGEN, NITROGEN, AND SULFUR*

The sample is burned in a quartz tube with oxygen injected into the helium carrier gas flow. The combustion is completed over copper oxide, and the excess of oxygen is removed and nitrogen oxides are reduced to nitrogen and sulfur trioxide to sulfur dioxide in a layer of metallic copper. The remaining combustion gases—nitrogen, carbon dioxide, water, and sulfur dioxide—are separated by gas chromatography and measured with a hot-wire detector and an integrator.

*See Dugan (1974, 1977), Pella and Colombo (1978), Kirsten (1979a).

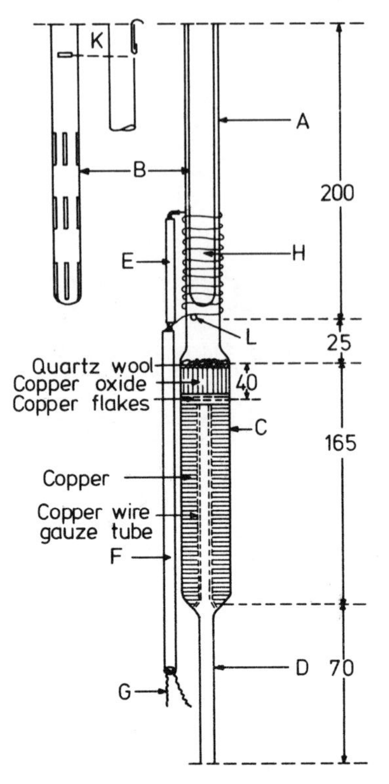

FIGURE 14. Reactor tube for CHNS determination. (A) Quartz tube, outer diameter 10.5 mm. (B) Inner tube that fits snugly into tube (A), with slits ground into it, length about 15–20 mm, width 1–1.5 mm. (C) Quartz tube, outer diameter 13 mm, wall thickness 1.5–2 mm. (D) Quartz tube, outer diameter 6 mm, inner 3 mm. (E) Pythagoras tube, outer diameter 3, inner 2 mm. (F) Pythagoras double-channel tube, outer diameter 5.8, inner 2 × 1.8 mm. (G) Kanthal A1 wire, diameter 0.8 mm. Two wires twinned together in the Pythagoras tubes, 35 turns of single wire around the flash combustion chamber (H). (K) Upper end of inner tube (B) with a horizontal slit and a nickel ribbon, 0.1 mm, which holds it hanging inside the outer tube (A). (L) Quartz wart on combustion tube that prevents the Kanthal windings from sliding down. The wire can be wound very tightly without risk of short-circuiting. Wind the copper wire gauze, 125-mm long, on the upper end of a 220-mm-long, 2.5-mm-diameter metal rod. Widen the lower end conically. Introduce rod and gauze from above into the reactor tube. Add the other fillings and draw out the metal rod.

It is important to use well-granulated materials for the tube fillings so that there is little resistance against the gas flow between the inlet membrane valve and the separation column. Dimensions in figure are millimeters.

APPARATUS

Use any of the two channels of the Carlo Erba 1106 analyzer or the CHN channel of the model 1104. Widen the bottom opening of the furnace with a grinding wheel mounted on a dentist's drill, so that the reactor tube (Figure 14) can be introduced from below, together with the Kanthal heater. If the thermocouple falls out during this operation, fix it again with fireproof cement or with a piece of Kanthal wire. Electrical contact between the Kanthal heater and the thermocouple must be avoided. Connect the reactor tube to the column with steel tubing (See Figure 18, Chapter 11) with an inner diameter of 0.5 mm and heated to 150–200°C.

Replace the 5-mL oxygen loop with a 10-mL loop and if necessary, introduce a delay loop as described in Chapter 9. Introduce a restricting capillary between the separation column and the detector as described in the Chapter 9. Prepare gas permeation tubes with sulfur dioxide as described in Figure 15. Keep the tubes in a freezer when not in use.

REAGENTS

Tungsten(VI) oxide. Boil tungsten(VI) oxide, Merck, Darmstadt No. 829, for 1 hour in 2% acetic acid solution. Filter and wash several times with hot redistilled water. Dry and heat to 1000°C in a quartz tube in a flow of oxygen for 3 hours.

Copper oxide from wire, Merck, (Darmstadt, FRG).

Copper prepared by reduction of the same copper oxide with pure, sulfur-free hydrogen.

Copper wire gauze from Engelsmann (Ludwigshafen am Rhein, FRG).

The reactor tube fillings should be as free as possible of such metals as lead, nickel, silver, and manganese, which form sulfates

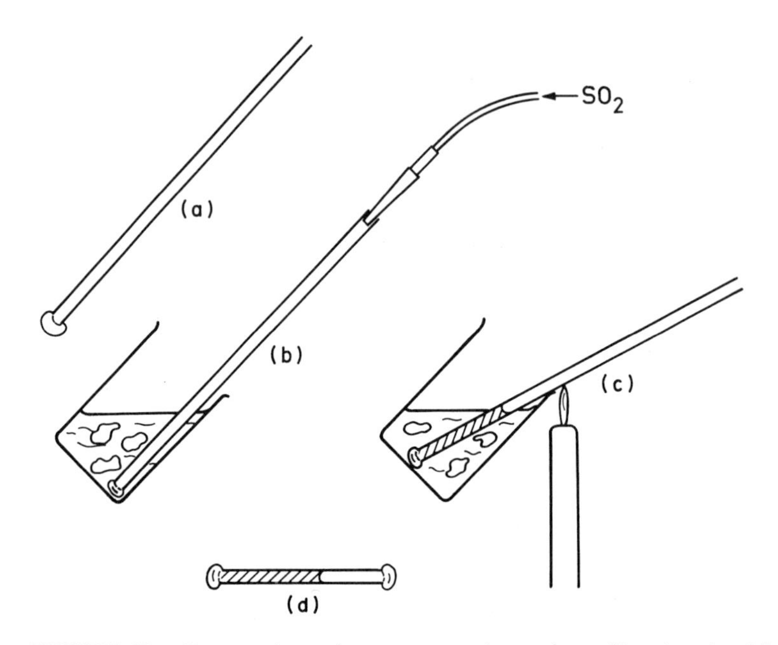

FIGURE 15. Preparation of gas permeation tubes. [Reprinted with permission from Kirsten (1979a). Copyright 1979 American Chemical Society.] (a) Shrink-Teflon tubing WTF-1241 from Tenntub Plastic Co., Inc. (Clifton Heights, Pennsylvania), outer diameter 4.1 mm, is heated at one end with a microburner, and the end is pressed together with a forceps. (b) A small beaker is charged with dry ice and acetone and clamped in an inclined position. (c) The tube is connected to a SO_2 tank with a tapered joint 5/20 and a pressure of about 1 kp/cm^2 is applied. SO_2 is now condensed in the tube. The tank is closed, the joint is removed, and the tube is heated with a microburner. When the tube melts, it is compressed with a forceps and, after cooling, the open end is cut off. (d) The resulting permeation tube. It is important to use double-layer shrink-teflon, otherwise it is difficult to get the tubes tight.

that are stable at much higher temperatures than copper sulfate and that therefore, can retain sulfur. The copper must be free from organic material. Such organic material can be deposited in the copper through reduction with impure hydrogen. It will cause apparent nitrogen blanks and low results of the other elements in the analysis of nitrogen-free samples through the water gas reaction. If this happens, eliminate the error with water vapor as described under CHN determination.

ADJUSTMENT OF APPARATUS

Set up and connect the apparatus as described by the manufacturer. Fill the reactor tube and attach the Kanthal heater as described in Figure 14. Check that the thermocouple is correctly in its groove in the furnace and introduce an empty CHN combustion tube, beak down, from above. Introduce the reactor tube from below so that the beak of the combustion tube is in the reactor tube. Move reactor tube and combustion tube upward until the reactor tube is in place. Take away the combustion tube and fix the reactor tube in position. Connect the heater to a variable transformer. Insert a thermocouple from below into the furnace so that its top is at the middle height of the copper filling. Tighten the bottom hole of the furnace with a refractory-fiber material to avoid air draft through it.

Measure the temperature of the flash compartment with a thermocouple inserted from above into it. Adjust it to 1030–1050°C. Adjust the temperature of the reduction zone to 730–760°C. The temperatures of the flash compartment and the reduction zone can be regulated with the regulator of the furnace temperature of the instrument and with the voltage of the Kanthal heater. In the model 1106 the instrument's regulator increases or decreases all temperatures in the furnace. Increasing the voltage of the Kanthal heater increases the temperature of the flash compartment and decreases that of the reduction zone and vice versa. In the model 1104 the two heaters work independently in the same direction.

When the working temperatures have been reached, run a few analyses with an easily combustible compound, like alanine or serine. Use a good excess of oxygen, (e.g., an oxygen pressure of 1.0 kp/cm^2 with a 10-mL loop). Use different times of sample inlet delay and observe the flashes to find out when the oxygen arrives at the flash compartment and when the last oxygen leaves it. Adjust the inlet delay time so that the sample falls into the combustion tube about 2–3 s after the arrival of the first oxygen. Do not use sulfur-containing compounds and do not use sulfur dioxide in the carrier gas in these analyses. When all the combustion conditions and all the chromatographic conditions are satisfactory, the apparatus is ready to work.

PROCEDURE

Weigh out the samples in tin capsules—preferably Lüdi 76 1308 83. Add 3–5 mg of tungsten(VI) oxide to each sample. Crimp the containers and compress their wings and place them in the 23 position sampler together with a gas permeation tube. If the 50 position sampler is used, the gas permeation tube is placed into a special doping tube in the gas line before the sampler; this tube is supplied by Carlo Erba for the oxygen determination.

Wait until a stable baseline is obtained and start analyzing according to the manual. Figure 16 shows a recorder trace of an analysis of phenylthiourea.

When the flash compartment of the reactor tube is filled, which happens after about 150–200 analyses, draw up the inner tube and replace it with a new one.

When the copper filling is consumed—oxidized to CuO—push

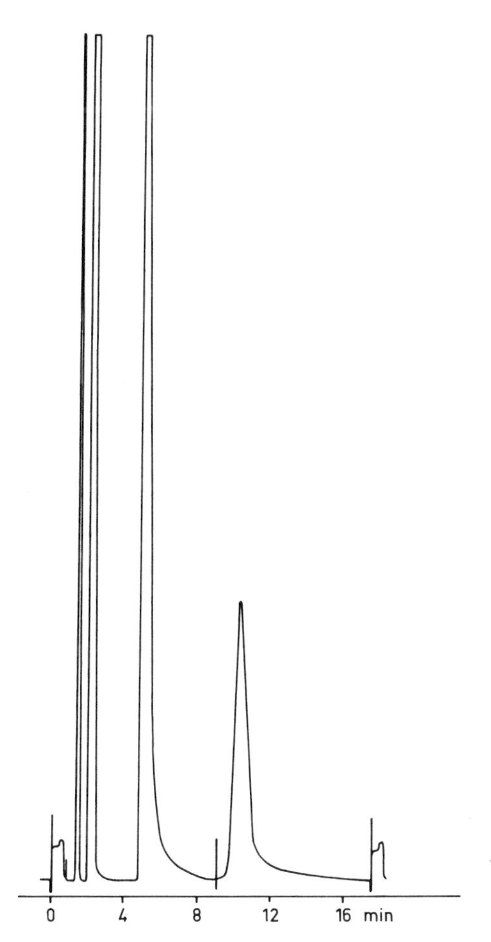

FIGURE 16. Recorder trace of CHNS determination. 539.8 μg of phenylthiourea were analyzed with the Carlo Erba 1106 instrument. Recorder sensitivity 1 mV. The peak at the start of the chromatogram was caused by a change of pressure when the injected oxygen was absorbed by the metallic copper. There is no such peak when the gas flow from the membrane valve to the copper filling is sufficiently free. The peak did not spoil the analytical result because the integrator was first started after the peak.

the standby button of the Model 1106. This will lower the temperature of the reduction zone to 500–600°C. When using the Model 1104, turn off the Kanthal heater. Turn off the detector current, close the helium outlet after the detector, disconnect the exit end of the reactor tube, and close the connection to the separation column. Insert a stainless steel capillary into the copper filling, and let a slow flow of very pure, sulfur-free hydrogen pass into it. When no more water vapor but only hydrogen comes out from the tube, remove the capillary. Place a dish below the tube, inject a few milliliters of water into the tube with a syringe, and brush with a pipe cleaner to wash out any salts. Take care not to push the pipe cleaner so high up that it is charred. Reconnect the tube and readjust the instrument to the working conditions.

When the reactor tube is replaced, the Kanthal heater can usually be left in place: Remove the sampler and the inner tube of the reactor tube and place an empty combustion tube, beak down, into the opening of the latter. Then draw out the reactor tube from below. The combustion tube follows the reactor tube down and keeps the Kanthal heater in place. Then introduce the new reactor tube from below as described earlier.

APPLICABILITY

The method is applicable to organic and inorganic compounds which can be decomposed through flash combustion with burning tin in the presence of tungsten(VI) oxide. In the author's laboratory the method has been in continuous routine use for the past 3 years, and it has not failed with any compound we received for analysis.

CHAPTER 11

AUTOMATIC SIMULTANEOUS DETERMINATION OF CARBON, HYDROGEN, NITROGEN, AND SULFUR OR TRACE SULFUR*

The sensitivity of the thermal conductivity detector is not sufficient for the determination of traces of sulfur, and the presence of traces of metal oxides, which form stable sulfates at higher temperatures than copper oxide in the reactor tube filling, interferes with the determination of traces of sulfur according to the CHNS determination method described earlier. A flame photometric de-

*See Sisti and Colombo (1978) and Kirsten (1981).

tector is therefore used for the measurement of traces of sulfur, and either the temperature of the reactor tube filling is increased or ultrapure copper which contains no interfering metals, is used.

APPARATUS

Use the same basic instrumentation as described earlier for the determination of CHNS with the separation system shown in Fig-

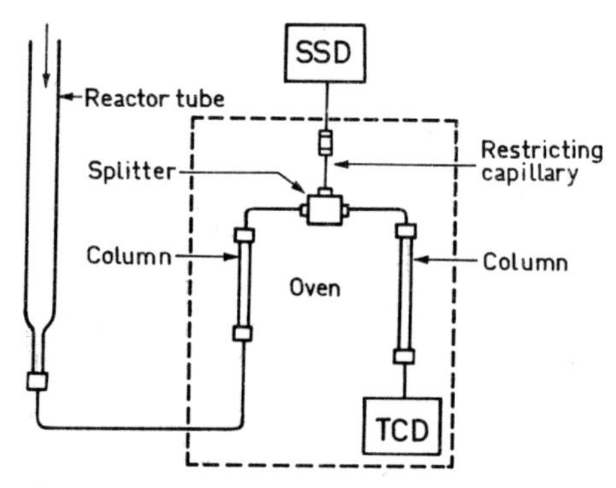

FIGURE 17. Separation system for CHNS or trace S. Left-hand column: 80 cm × 5 mm, glass, filled with Porapak QS; right-hand column: 120 cm × 5 mm, glass, with the same filling. The splitter is an ordinary Carlo Erba T-piece. The restricting capillary is a steel capillary, 0.2 mm inner diameter, from Carlo Erba, which is compressed with a small vise or a pair of very strong pliers to give the desired flow resistance. (A better method is to introduce a fine wire into the capillary as described in Chapter 9.) The flow through the thermal conductivity detector (TCD) should be 30 mL/min and that through the flame photometric detector (SSD) should be about 0.5 mL/min. When the flow rate is adjusted with capillary and columns at room temperature, it decreases 25–30% when capillary and columns are heated to working temperature. The connections between reactor tube, left-hand column, splitter, and SSD are made with the flexible connections shown in Figure 18a, and from the splitter to the right-hand column and the detector with those shown in Figure 18b.

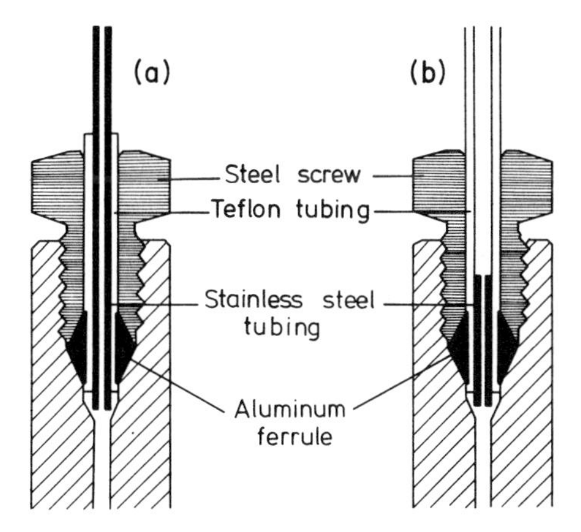

FIGURE 18. Flexible connections. Connectors and ferrules from Carlo Erba. Stainless steel syringe canule tubing, outer diameter 1.0 mm, inner 0.5 mm. Teflon tubing, outer diameter 2.0 mm, inner 1.0 mm.

ures 17 and 18. Use the Carlo Erba flame photometric detector SSD 250 together with a separate integrator and prepare the gas permeation tube according to Figure 19.

REAGENTS

If the copper filling of the reactor tube is to be held at a temperature of 850–900°C, use the same reagents as described earlier for the determination of CHNS. If it is to be held at 740–760°C, use ultrapure copper wire, 99.999%, 1 mm diameter, available from Koch-Light Laboratories Ltd. (Colnbrook Bucks, England). Cut the wire into 3–4-mm-long pieces and place them loosely into a quartz tube. Oxidize and reduce them alternately two or three times thoroughly with oxygen and hydrogen to obtain sufficient activity and volume. Carry out the reduction slowly at about 300°C with very pure hydrogen, free from traces of organic and sulfur-containing com-

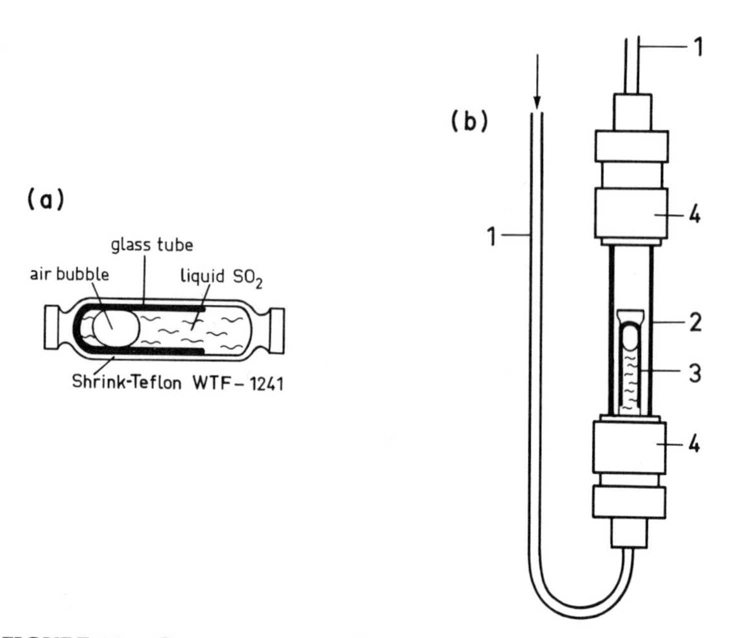

FIGURE 19. Gas permeation tube and its mounting for trace sulfur determination. (a) A glass tube with one closed end is fitted snugly into shrink-Teflon tubing (outer diameter ~4.3 mm) and is shrunk into it hermetically in the same manner as that shown in Figure 15a. It is then filled with sulfur dioxide and sealed off in the same manner as that shown in Figure 15. The length of plastic tube that is open for diffusion is about 3 mm. The quantity of sulfur dioxide liberated can be estimated roughly as about 0.5 μg/min. (b) The gas permeation tube is mounted into the carrier gas line—2-mm steel tubing (1)—in thick-walled brass tubing (2) with Carlo Erba connectors (4). The tube with mounting is heat-insulated in a Dewar flask or in a heavy metal container.

pounds. Metallic copper that has not been sufficiently oxidized and therefore is too massive will swell in the reactor tube and break it.

Tungsten(VI) oxide. Heat tungsten(VI) oxide, 99.999%, No. 20,478 from Aldrich Chemical Co. Inc., P.O. Box 355, Milwaukee, Wisconsin 53201, for 3 hours in a quartz tube to 1000°C in a flow of oxygen.

ADJUSTMENT OF APPARATUS

Assemble the instrument as described in the legends of Figures 17–19. Run a few analyses as described earlier under CHNS determination to determine the optimal time for the sample inlet. Introduce the gas permeation tube and run a few analyses with samples containing carbon, hydrogen, and nitrogen and a few percent sulfur for the adjustment of the chromatographic conditions and the integrator programs. Instruct the trace sulfur integrator to start integrating about 30 s to 1 min before the appearance of the sulfur dioxide peak.

It is very important that not only the carrier gas flow rate but also the flow rates of the hydrogen and oxygen to the detector are quite constant. Variations do not affect the baseline, but they affect the response of the detector and cause inaccurate results.

PROCEDURE

Carry out the analyses in the same manner as described earlier under CHNS determination using the Aldrich tungsten(VI) oxide. Do not run a trace determination right after a sample that contains a large amount of sulfur because the last traces of sulfur from such a sample are not completely eluted during the analytical cycle and some remaining sulfur might cause high results in a trace determination.

Trace Determination of Sulfur Alone

When only traces of sulfur are to be determined, the signals from the hot-wire detector can be disregarded. This makes it possible to shorten the analytical cycle. The next analysis can be started as soon as the trace sulfur peak has emerged through the flame photometric detector.

CHAPTER 12

AUTOMATIC MICRO AND TRACE DETERMINATION OF NITROGEN*

The samples are encapsulated and burned, and gases are passed through the reduction tube in the same manner as that in the CHN determination method. Water and carbon dioxide are then absorbed in magnesium perchlorate and alkali, and the nitrogen passes through a short separation column into a hot-wire detector. Combustion tube, reduction tube, absorption tubes, and sampler are large enough to allow the analysis of samples containing up to 50 mg of organic material. One analysis takes 5 min. The sampler can be loaded continuously while the instrument is running, and about 100 analyses can be carried out in a working day. The modified sampler (see Fig. 23), can then be loaded with further 98 or more samples for unattended analysis overnight.

*See Colombo and Giazzi (1982) and Kirsten and Hesselius (1983).

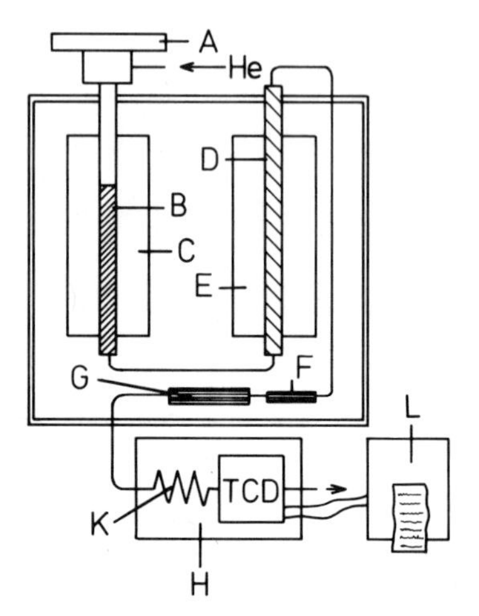

FIGURE 20. Layout of ANA 1400. [From Kirsten and Hesselius, (1983).] (A) Sampler with drum with 50 holes. (B) Combustion tube in (C) furnace at 1030°C. (D) Reduction tube in (E) furnace at 650°C. (F) Water absorption tube. (G) Carbon dioxide absorption tube. (H) Oven. (K) Separation column. (TCD) Thermal conductivity detector. (L) Integrator-printer.

The analysis starts with the injection of 25 mL of oxygen into the helium carrier gas (He). Just before the oxygen reaches the combustion chamber the sampler drops a sample, which is burned by the nickel oxide and the oxygen in the combustion tube. The gases pass into the reduction tube (D), which is filled with metallic copper. The copper absorbs the excess of oxygen and reduces nitrogen oxides to nitrogen. The water absorption tube (F) then removes the water, and the tube (G) removes the carbon dioxide. The nitrogen then passes with the carrier gas through the column (K) and through the thermal conductivity detector (TCD). The signal of the nitrogen peak is integrated and the result printed out by the integrator-printer (L).

APPARATUS

(a) Automatic Nitrogen Analyzer, ANA 1400, from Carlo Erba Strumentazione (Milano, Italy). See Figure 20.

(b) Equipment for sample encapsulation according to Pella and Colombo (1973), modified as shown in Figure 20.

(c) Equipment for encapsulation of volatile and nonvolatile liquids.

MODIFICATIONS

The instrument works well as supplied. The following modifications are, however, recommended: Fill the combustion tube as shown in Figure 21. Shorten the heating jacket between the combustion tube and the reduction tube to 170 mm and its cartridge to 160 mm. Make the connection as shown in Figure 22.

Omit the water absorption tube and use a 345-mm-long carbon dioxide absorption tube as shown in Figure 22. The tube can be bypassed with a standby shunt, as shown in the figure, in order to avoid too rapid consumption of the magnesium perchlorate in the standby position.

Shorten the separation column to 50 cm and introduce a short restricting capillary between it and the detector. The capillary is a restricting capillary with a 0.25-mm inner diameter. Adjust its flow resistance as described in Chapter 9 so that the flow rate through the system at room temperature is about 60 mL/min at a helium overpressure of 0.5–0.6 kp/cm^2. The same flow rate will then be obtained at a working temperature with an overpressure of about 0.8–0.9 kp/cm^2.

Use the following analytical program: Oxygen in at 2 s. Sample in at 35 s. Oxygen inlet stop at 1 min 3 s. Sampler back at 1 min 8 s.

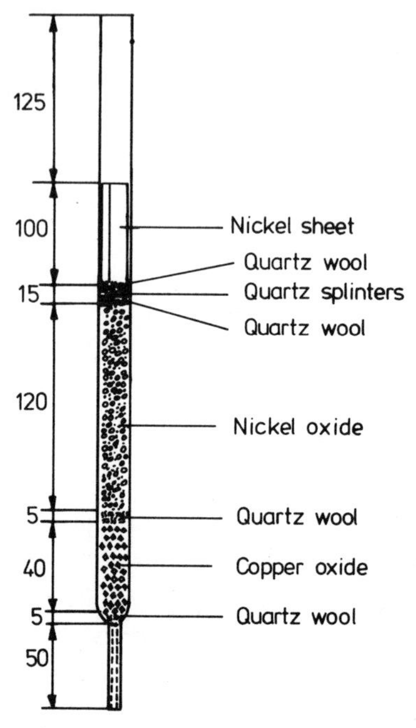

FIGURE 21. Combustion reactor for ANA 1400. [From Kirsten and Hesselius (1983).] Quartz combustion tube, outer diameter 16.5–17.0 mm, wall thickness 1.5–1.75 mm, preferably of opaque quartz, which has a longer lifetime than clear quartz. Beak, outer diameter 6.0 mm, inner diameter 3 mm. The nickel sheet is 0.1-mm nickel plate, well overlapping and snugly fitting into the tube. There must be no fine dust in the tube fillings, and the quartz wool must not be compressed too tightly, which would restrict the free gas flow. The copper oxide must be introduced into the tube through another tube inserted into it to prevent copper oxide particles from staying in the upper parts of the tube. Dimensions are millimeters.

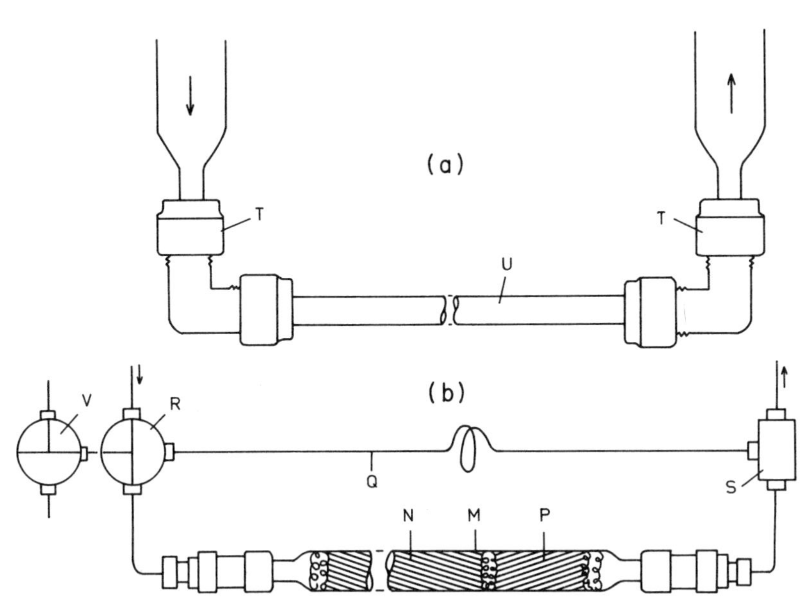

FIGURE 22. Modifications of connections in the ANA 1400. [From Kirsten and Hesselius (1983).] (a:) Connection between combustion tube and reduction tube. (T) Swagelok union elbows, ¼ in. (U) Stainless steel column tubing, outer diameter 6 mm, length 200 mm, filled with quartz splinters between loose wads of quartz wool. Tube (U) is tightened with brass ferrules, the quartz tubes are tightened with rubber O-rings. (b:) (R) Three-way stopcock B-41XS2 from Whitey, (Oakland, California), mounted in the lower perforated plate at the left side of the instrument. (Q) Capillary 0.2 mm. (S) T-connection, Carlo Erba. (V) Stopcock (R) in the standby position. (M) Absorption tube, length 345 mm, outer diameter 17 mm. (N) Carbon dioxide absorbent, length of filling 200 mm. (P) Magnesium perchlorate, water absorbent. The narrow ends of tube (M) are left empty.

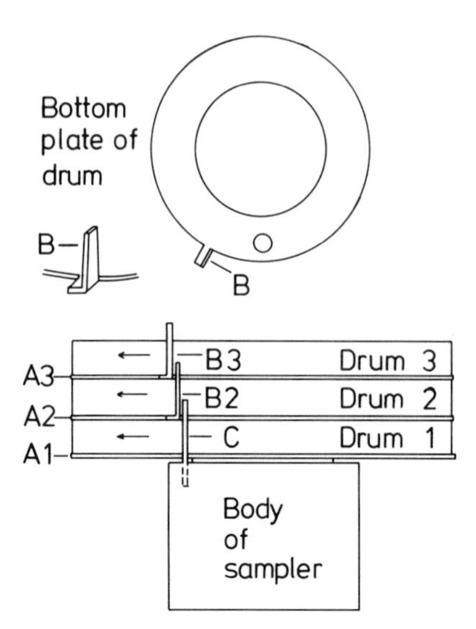

FIGURE 23. Multiple drum sampler [From Kirsten and Hesselius (1983).] Remove the ring around the drum of the Carlo Erba 50-position sampler. Remove the midhandles of the drums and fix two short, vertical rods opposite each other in the upper side of the drums, which fit into corresponding holes drilled in their bottoms, so that the drums are stable when placed on top of each other with their sample holes aligned.

Remove the rings, which hold bottom plates (A2) and (A3) to their drums, and make new bottom plates of Plexiglas with stop-noses (B2) and (B3). Fix the stop-rod (C) in the body of the sampler.

Load all drums with samples and place them on each other as shown. When the instrument starts, all drums move around stepwise with their bottom plates, except bottom plate (A1). Drum 1 discharges its samples, one after one, in the usual way. After one turn, stop-nose (B2) meets the stop-rod (C), and the bottom plate stops. All drums and bottom plate (A3) continue to rotate. Drum 2 discharges its samples, one after the other, through the hole in bottom plate (A2) and through the sample holes in drum 1. Then the stop-nose of drum 3 meets the stop-nose of drum 2. Bottom plate (A3) stops and drum 3 discharges its samples, and so on.

Theoretically, a large number of drums can be stacked on each other in this way but so far we have only used two drums.

Stop-rod (C) and stop-noses (B) must be placed so that the openings in the bottom plates are exactly over each other and over the openings in the drums and in the sampler body when the bottom plates stop. Fine adjustments can be made by bending the rod and the noses.

Tape a ring of filter paper to the underside of the bottom plate of every drum (B) with double-stick tape to increase its friction with the drum below.

Autozero on at 1 min 4 s. Integrator on at 1 min 28 s. Autozero off at 1 min 36 s. Integrator off and memory on and 4 min 35 s. Print at 4 min 50 s. End of cycle at 5 min. It is necessary to cut off a piece of the comb 4 of the programmer with a sharp knife in order to get the longer integration time.

With these modifications the sample falls into the combustion tube just before the arrival of oxygen, which minimizes the risk of formation of nitrogen oxides during combustion, and the nitrogen peak comes earlier; also, very large peaks are completely eluted and integrated and do not interfere with a following blank or trace determination.

In order to increase the capacity of the instrument for overnight runs, the sampler can be modified as shown in Figure 23.

The instrument is provided with a peak sensor, which turns off the analyses when a small peak or no peak at all is obtained in an analysis. It is presumed then that all analyses have been done. This makes it impossible, however, to carry out blank and trace determinations in an automatic run. It is therefore recommended that this peak sensor be inactivated and that the instrument be provided with a time switch that turns off the analytical cycle and the recorder after the run.

REAGENTS

Granulated nickel oxide (see Chapter 9).

Copper wire for elemental analysis.

Quartz splinters, 1.0–1.5 mm.

Carbon dioxide absorbent: Mix 3 volumes of sodium asbestos with 7 volumes of *dry* soda lime. Do not use fine particles or dust. The soda lime prevents the filling from clogging.

Magnesium perchlorate for drying. Do not use fine particles or dust.

Quartz wool.

Calibration substances (e.g., phenacetin, benzimidazole, urea, melamine).

Nitrogen-free test substances (e.g., sucrose for the analysis of ordinary solids, liquid paraffin for the analysis of oils and other hydrocarbons).

ADJUSTMENT OF APPARATUS

Adjust the apparatus according to the manual, except for the modifications described earlier.

Use the following running conditions: temperature of combustion furnace, 1030°C; temperature of reduction furnace 650°C; temperature of oven 80°C; gas flow rate, 60 mL/min; detector current, 120 mA; oxygen pressure, usually 1 kp/cm^2.

ENCAPSULATION OF SAMPLES AND LOADING OF DRUMS

It is *very important* to encapsulate all samples so that no jamming can occur in the drum or in the body of the sampler. With the following procedures many thousands of analyses have been carried out in the author's laboratory without a single case of jamming.

Encapsulate *stable solids* as shown in Figure 24.

Encapsulate *hygroscopic solids* in the same manner in a glass bell under dry gas, as shown in Figure 7, and weigh quickly. Moisture absorbed after the weighing does not affect the results.

Introduce *liquids* into smooth tin capsules with a height of 6.0

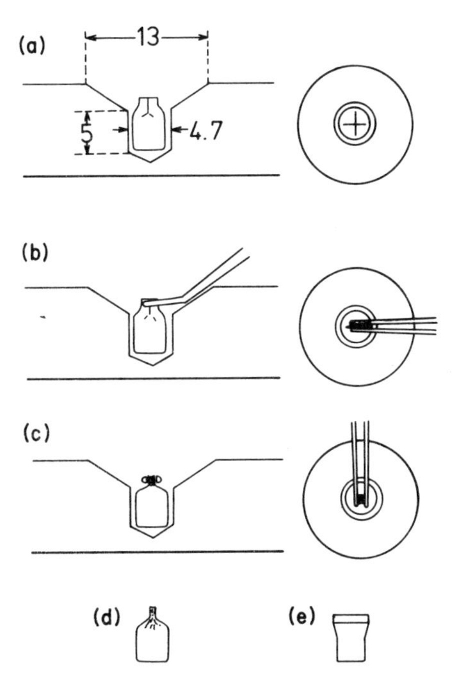

FIGURE 24. Encapsulation of samples. [From Kirsten and Hesselius (1983).] Capsule: pressed tin capsule, height 6 mm, inner diameter 4 mm, weight 18 mg, Lüdi No. 76 1305 53. (a) Carlo Erba crimping rod and modified block with hole. Crimp the capsule with the sample as shown. Compress the wings with forceps (b) first in one direction and then (c) in the perpendicular direction. Grasp the upper end of the capsule as shown in (c), lift it out of the hole, and place it into the drum of the sampler. *Never grasp it in any other way!* This might cause deformation of the capsule and jamming in the sampler. (d) The closed capsule. Note: When filling the capsule, have it standing on a grounded plate in front of the balance, not in the crimping block, and weigh it before the crimping. (e) A closed capsule for liquids. Dimensions are millimeters.

mm, an inner diameter of 2.9 mm, and a weight of 34 mg [No. 84 0180 41 from Lüdi, Metallwarenfabriken (Flawil, Switzerland)] with a syringe. Avoid drawing air bubbles into the syringe! Nonvolatile liquids can also be introduced with a pipette or a metal wire. Fill the capsules completely. Seal them then by cold welding with pliers or by hot welding with a spot welder (see Chapter 5). The upper layers of the liquid, from which fractional evaporation may have occurred, are thus squeezed out from the capsules, leaving the remaining liquid with its original composition.

When the capsules are sealed at their very tops, their content is about 20 μL. Smaller samples are obtained by sealing the capsules nearer their bottoms. No flushing with helium or oxygen is required because the capsules are completely filled.

Wash the capsules, if necessary, with acetone or another suitable solvent.

Encapsulate *other difficult materials* as described in Chapter 5.

When the apparatus can be attended, use only one drum in the sampler and run the analyses as described in the manual of the instrument. When many analyses are to be run overnight with the sampler, (Figure 23), fill up drum 1, put on the empty drum 2, draw its bottom plate to the start position, and fill it with samples, starting with the first place after the hole in the bottom plate, regardless of the numbering on the drum, and so on.

When loading drums not situated on the sampler, clamp the bottom plate to the drum with a trouser clip, as used for bicycling. Put the loaded drum on the sampler and draw out the clip.

RUNNING THE ANALYSES

Follow the manual. It is, however, strongly recommended that a calibration run and a blank run with a nitrogen-free substance be carried out after, say, every 15 unknowns. In this manner the method is always kept under good supervision and the return of characteristic peaks after regular intervals makes it easier to correlate samples and printed results in case some irregularity should occur in a series of analyses.

When the nitrogen-free substance begins to give unacceptably high blanks, replace the nickel oxide in the combustion tube with a fresh charge.

About 300–400 samples containing about 15 mg of organic material can be analyzed with good accuracy with a nickel oxide filling. The relationship between the size of the samples and the number of analyses that can be run with one filling is not linear. Many more small samples can be analyzed.

When the flash combustion part of the combustion tube is filled with residues from the samples, close the carrier gas outlet, turn off the detector current and the helium flow, take out the combustion tube, and scratch out the residues and most of the quartz splinters with a steel hook. Add new quartz splinters and quartz wool and replace the tube. Turn on the helium flow, open the carrier gas outlet, and after a while turn on the detector current.

Follow the same procedure also when replacing the nickel oxide, the reduction tube, and the absorption tube. Make all such replacements quickly, preferably with prefilled or preemptied tubes as needed.

When a new combustion tube is introduced, or a tube with a new nickel oxide filling, burn out the tube first by triggering a few analyses before connecting it to the reduction tube.

Reduce consumed reduction tubes as described for CHN analysis in Chapter 9.

Avoid analyzing substances that liberate large amounts of nitrogen oxides in the combustion (e.g., inorganic nitrates, because of the reasons discussed in Chapter 13).

APPLICABILITY

An important application of the method is the determination of traces of nitrogen in oils. The samples are encapsulated as described earlier under encapsulation of liquids. Since the samples have not been flushed under a nitrogen-free gas, they contain dissolved atmospheric nitrogen. The solubility of nitrogen in liquid paraffin oil is about 0.009%, calculated according to Ridenour *et al.* (1954). This is compensated for automatically by using ultrapure nitrogen-free liquid paraffin as a blank and subtracting the blank value.

The method is also applicable for the analysis of technical, industrial, and agricultural raw materials and products. Grain, peat, chipped wood, and food products can be ground finely in a very short time with a Retsch ZM1 centifugal mill with 24 wings and a 0.08-mm sieve. The flour is fine enough to give reproducible results with 2–3-mg samples. Samples of about 15 mg are usually used for these analyses. Compare Kirsten and Hesselius (1983).

CHAPTER 13

ADDITIONAL REMARKS ON CHN, CHNS, CHNS TRACE S, AND N DETERMINATION METHODS

The basic instrument and the CHN method were described by Poy (1970). The methods were studied and improved by Pella and Colombo (1972, 1973, 1978) and by Sisti and Colombo (1978). The described procedures are modifications by Kirsten (1979a, 1983).

REACTOR TUBES

Quartz tubing is strongly attacked by burning tin capsules, by hot copper oxide, and by hot cobalt oxide–silver. Exposed parts of quartz combustion tubes are, therefore, protected with nickel sheets, as shown in Figure 13. Opaque quartz is considerably more resistant than clear quartz, and thick-walled quartz is much more resistant than thin-walled (Kirsten, 1952). Even with nickel protection the lifetime of a quartz tube is seldom longer than a few weeks. Nickel tubes can be used for many months.

Nickel tubes and nickel protection sheets cannot be used when sulfur is also to be determined because nickel sulfate, which is stable at a much higher temperature than copper sulfate and which would cause retention of sulfur in the tube, is formed. Disposable inner tubes of quartz are, therefore, used in the CHNS methods.

RETENTION OF NITROGEN

In the Dumas determination of nitrogen a retention of nitrogen, particularly of nitrogen linked to oxygen, was observed when free oxygen was used in the combustion (Kirsten, 1952). The same retention of nitrogen was observed in the CHN method when the combustion tube was held at 940°C. Correct results were, however, obtained at 1030°C–1050°C, the temperature recommended by Pella and Colombo (1973). Kirsten and Hesselius (1983) found that nitrogen oxides, which are formed in the combustion, are completely reduced in the copper layer of the reduction tube but that about 5.6% of their nitrogen is retained in the beginning layer of the copper. When this copper layer is oxidized in following analyses, the retained nitrogen is liberated, which causes slightly high results. The combustion should therefore be carried out in such a manner that no or very little nitrogen oxide is formed. At the high temperature the flash occurs so violently that a compact reducing cloud is momentarily formed, in which most nitrogen linked to oxygen is reduced.

Inorganic nitrates or other compounds containing nitrogen

linked to oxygen and not containing organic material cannot be analyzed with this method.

INLET DELAY TIME

The correct timing of the inlet delay and a sufficient excess of oxygen are important in all methods involving the determination of sulfur. If there is a deficiency of oxygen, the recorder traces still look fine and the CHN results are correct, but the sulfur results are erroneous, probably because sulfide is formed.

TEMPERATURE OF COPPER FILLING

At temperatures above 775°C copper reacts with oxygen and forms Cu_2O. Below this temperature it forms CuO. This means that the capacity—and the lifetime—of the tube filling at the low temperature is twice that of the filling at the higher temperature. Furthermore, at the higher temperature the copper—or rather the copper oxide formed from it—attacks the quartz tube very strongly and also the filling sinters. It is therefore not possible to reduce and reuse a copper filling that has been kept at the high temperature as many times as one that has been kept at a lower temperature.

However, most specimens of copper or copper oxide contain traces of metals that form sulfates, which are stable at higher temperatures than copper sulfate and which, therefore, retain sulfur.

In the ordinary CHNS determination the reduction tube can be kept at low temperature, and it is sufficient to block the active metals with the sulfur dioxide doping through the gas permeation tube.

When traces of sulfur must be determined, this is not sufficient. Either the reduction filling must be kept at 850–900°C, where also most of the more stable sulfates are decomposed, or an ultrapure copper must be used, which is very expensive and tedious to get into the applicable shape.

It is also possible to use an electron capture detector instead of the flame photometric detector for the trace sulfur determination. It gives a straight calibration curve. Two different brands of detectors that we tried, however, gave rather high noise so that their highest sensitivity could not be used. We had to split off a larger part of the combustion gas through them. They were also not as stable as the flame photometric detectors.

CHAPTER 14

AUTOMATIC DETERMINATION OF OXYGEN*

The sample is pyrolyzed over carbon in a flow of helium doped with chloropentane vapor. The oxygen of the samples is thus converted to carbon monoxide, which is separated from accompanying hydrogen and nitrogen in a chromatographic column and measured with a hot-wire detector and an integrator.

APPARATUS

The oxygen channel of the Carlo Erba elemental analyzer is used with the additional accessories shown in Figures 25 and 26. The

*See Poy (1970), Pella and Colombo (1972), Pella and Andreoni (1976), and Kirsten (1977, 1978a).

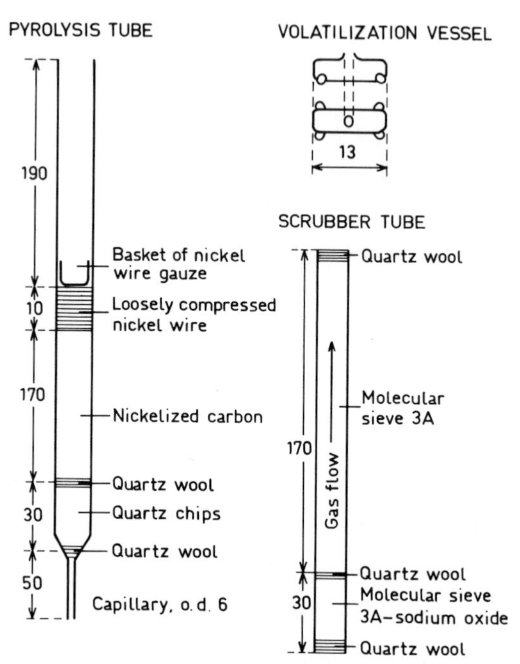

FIGURE 25. Pyrolysis tube, scrubber, and volatilization vessel. [From Kirsten (1978a).] Use the widest possible pyrolysis tube. The lifetime of the tube increases very strongly with the volume of the carbon filling. It can be advantageous to widen the bottom opening of the furnace by grinding it with a diamond drill or to take out the bottom gable of the furnace completely before use. The filled part of the tube can then be made wider than the regular 10.5–11 mm. Prepare the nickel wire gauze basket by folding tightly woven wire gauze, wire diameter 0.15 mm, around the end of a metal rod. The volume of the volatilization vessel should be about 0.3 mL and the diameter of its orifice should be about 1.5 mm.

Mix equal volumes of conditioned molecular sieve 3A and sodium oxide, carefully protecting both from moisture. Fill the scrubber as shown. Dimensions are millimeters.

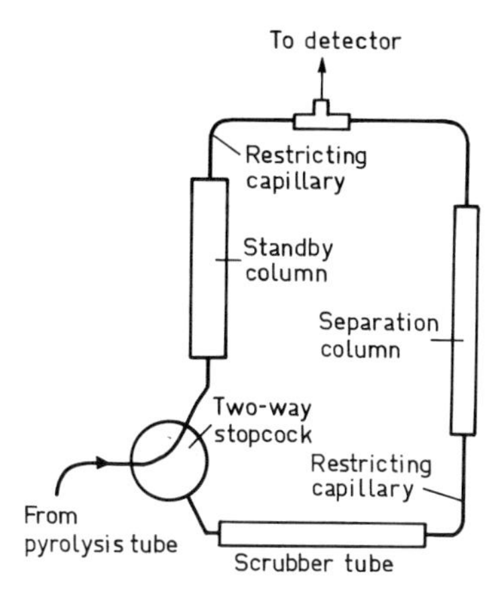

FIGURE 26. Bypass system. [From Kirsten (1978a).] Separation column stainless steel, 40 cm × 5 mm, molecular sieve 5A. Standby column stainless steel, 20 cm × 8 mm, molecular sieve 5A. Restricting capillaries stainless steel 60 mm × 0.2 mm. Two-way stopcock, Whitey B41XS2 (Oakland, California).

scrubber and the separation column with or without bypass are mounted outside of the oven in model 1104 and below the furnaces in model 1106. The two-way stopcock is conveniently mounted in a hole drilled into the wall of the instrument.

The original instrument is supplied with a longer separation column mounted in the oven. This column lasts considerably longer than the short, outside-mounted column, but it is more difficult to exchange and to refill it.

REAGENTS

Nickelized carbon, *ca.* 20% Ni: Dry Degussa CK3 or another good pelletized carbon black at 200°C overnight. Dissolve 200 g of nickel nitrate, $Ni(NO_3)_2 \cdot 6H_2O$, in water and dilute the solution to

300 mL. Pour the solution on 80 g of the dry carbon black and mix well. Transfer the slurry to a filter paper on a Büchner funnel. Transfer carbon residues with filtrate and suck nearly dry. Transfer the carbon into a wide quartz tube, heat very slowly under hydrogen to 1100°C, and let cool to room temperature under hydrogen. Keep in a glass bottle with a rubber stopper.

Nickel wire: 0.07 mm, from Alloy Wire Co. (Cradley Heath, Warley, U.K.).

Molecular sieves: 5A for standby column, Merck Darmstadt 5705; 5A for separation column, 80–100 mesh, for gas chromatography, W.G.A. International Ltd. (London); 3A for scrubber, Merck 5704 (Darmstadt, FRG) ground and sieved to particle size between 200 and 800 μm. The ground and sieved molecular sieve 5A, Merck 5705, can also be used for the separation column. It is much less expensive than the special material for gas chromatography. Condition the molecular sieves at 300–350°C according to the description of the manufacturer and keep them carefully protected from moisture.

Sodium oxide: Merck 806558 (Darmstadt, FRG) for synthesis.

Doping solution: 1 volume of 1-chloropentane, Fluka No. 25770 (Switzerland), mixed with 1 volume of toluene, Merck 8331 (Darmstadt, FRG).

Capsules: Silver capsules, Lüdi 76 1308 82 for solid samples and tin capsules, Lüdi 84 0176 01 for liquids. Wash the capsules with acetone and alcohol and dry them at 75°C in air. The silver capsules contain no detectable amounts of oxygen after washing and drying; the tin capsules contain about 0.3 μg per capsule.

ADJUSTMENT OF APPARATUS

Set the temperature of the pyrolysis furnace to 1040°C and the gas flow rate to 50 mL/min. Keep the two-way stopcock always in the standby position, except when oxygen determinations are carried out. Adjust the integrator so that it takes the baseline just before the elution of the carbon monoxide peak.

PROCEDURE

Encapsulate the samples. Hermetically sealed capsules must not contain air. Seal them under inert gas or fill them completely as described in Chapter 5. Place them into the sampler drum together with test samples. Fill the volatilization vessel with 0.15 mL of doping solution and place it in the groove of the drum. Switch the two-way stopcock to the running position, let the baseline stabilize, and start analyzing, as described in the manual of the instrument. After the run, switch back the two-way stopcock to the standby position. When the basket in the pyrolysis tube is full of residues, remove the sampler, dip a 3-mm stainless steel rod into the basket, wait a few seconds, and draw up the rod. The silver will solidify, cooled by the rod, and silver and basket are removed along with the rod. If this does not work turn off the detector current, draw up the tube, close its lower end with a rubber cap, and scratch out residues and basket with a steel hook. Insert a new basket and replace tube and sampler.

Analyze metal salts and oxides in the same way. In some cases it might be preferable to use pure chloropentane instead of the doping solution in order to provide for a rapid decomposition of the sample, and in some cases it might be necessary to use a longer cycle time (cf. Figure 27).

Refill the scrubber after about 200–300 analyses. When the efficiency of the separation column decreases, decrease the rate of the gas flow. Refil both scrubber and column when the separation becomes unsatisfactory.

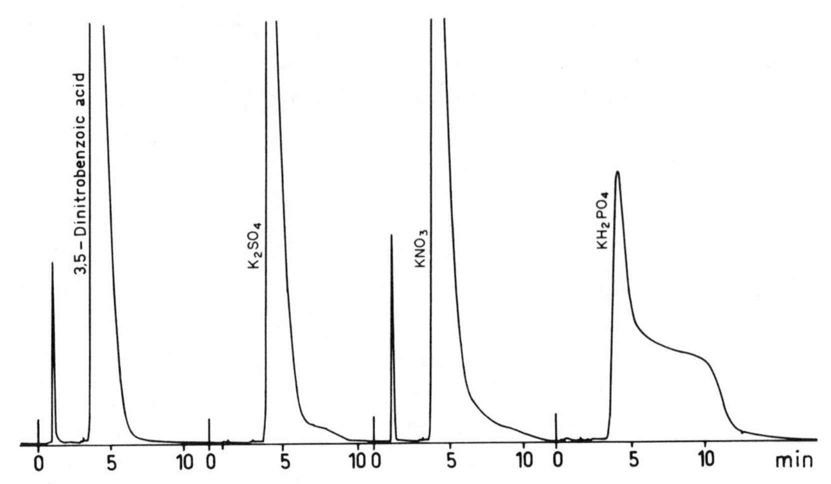

FIGURE 27. Recorder traces obtained with different compounds with the Carlo Erba 1104 instrument. Nitrogen and hydrogen peaks attenuated ¼. Pure 1-chloropentane was used for the doping. [From Kirsten (1978a).]

APPLICABILITY

The method is used for the analysis of pure organic compounds as well as for industrial and agricultural products. Large samples can be used for trace determinations (e.g., several milligrams of mineral oil).

Fluorine compounds give peaks that are measured as oxygen, even when quartz-free reactor and scrubber tubes are used (Kirsten 1978a).

The oxides of several metals and nonmetals are so stable that they are decomposed only incompletely or not at all (e.g., the oxides of aluminum, boron, silicon, vanadium). Most alkali salts and the salts of alkaline earths are decomposed and give correct results when pure chloropentane is used for the doping. They are converted to chlorides under these conditions. The conversion requires a somewhat longer time for some salts (cf. Figure 27). A somewhat longer cycle time must therefore be used for these analyses.

Since water is composed of 89% oxygen it is important to make sure that the samples are dry.

DISCUSSION OF THE OXYGEN DETERMINATION METHOD

The oxygen determination method is essentially that described by Pella and Colombo (1972) and Pella and Andreoni (1976), which was based on the pyrolysis reactions used in the methods of Schütze (1939) and Unterzaucher (1940), and gas chromatographic separation and measurement of the formed carbon monoxide with a thermal conductivity detector as described by Poy (1970).

The sample is pyrolized in a flow of helium, and the products of the pyrolysis are passed over a carbon contact, which converts all oxygen compounds to carbon monoxide.

Unterzaucher (1940) had stated that sample and carbon contact must be held at 1120°C to provide for a complete recovery of the sample's oxygen. Oita and Conway (1954) stated that a satisfactory conversion of oxygen to carbon monoxide can be obtained at 900°C if platinized carbon is used as a catalyst, and Terentev *et al.* (1963) found that nickelized carbon could be used as well. Thereafter many analysts used temperatures lower than 1120°C. However, we found that although the conversion of released oxygen in the catalyst is complete, many materials do not release all their oxygen at these lower temperatures, which caused low results and blanks and other memory effects (Kirsten 1978a). We found also that doping the carrier gas with chlorohydrocarbon vapor enhances the liberation of the oxygen from the sample, which makes it possible to use a lower temperature for the pyrolysis than 1120°C, which is quite harmful to the quartz tube and its filling. Temperatures of 1020–1050°C are quite convenient.

Some volatile compounds, probably mostly nickel chloride, are formed in the reaction. They would slowly pass into the separation column and decrease its efficiency. They are held back by the scrubber. The lifetime of the scrubber filling is increased by the bypass system (Figure 26). However, without the bypass system the scrubber filling can be used for at least 12 drums of samples with the 23-sampler, when pure chloropentane is used for the doping, and many more if a mixture with less chlorine is used. And it takes only a few minutes to renew the fillings of the scrubber and the separation column.

CHAPTER 15

ULTRAMICRO DETERMINATION OF SULFUR*

The apparatus shown in Figures 28 and 29 is used.

The samples, moistened with orthophosphoric acid, are burned in the combustion tube (L) (Figure 28) in a flow of oxygen. The combustion gases pass through orifice U into chamber M into an excess of hydrogen at 1200°C, which enters through a side tube. All sulfur is hydrogenated to hydrogen sulfide, which is absorbed in a flask with zinc acetate solution and measured by spectrophotometry after reaction to ethylene blue.

*See Kirsten (1950, 1962, 1978a) and Rees *et al.* (1971).

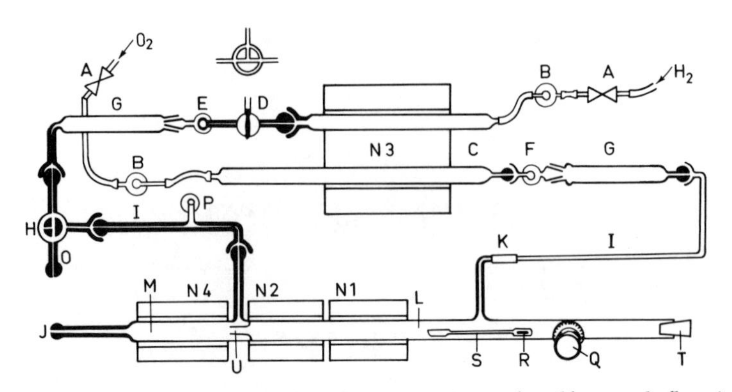

FIGURE 28. Apparatus for determination of sulfur and fluorine. [Reprinted with permission from Kirsten and Shah (1975). Copyright 1975 American Chemical Society.] (A) Needle valves for fine adjustment of gas flow rates. (B) Rotameters. (C) Quartz tubes in furnace (N3) at 950°C. (D) Three-way stopcock of scrubber (E) (cf. Figure 24). (F) Scrubber (cf. Figure 24). (G) Soda lime tubes. (H) Three-way stopcock. (I) Glass capillary connections. (J) Joint for absorption flask. (K) Tubing of silicone rubber. (L) Combustion tube, quartz, inner diameter 8 mm and outer diameter 11 mm. (M) Hydrogenation chamber, sealed on combustion tube, inner diameter 9 mm, outer diameter 12 mm. (N1, N2), split-type combustion furnaces, temperature 700°C for sulfur, 900° for fluorine, lengths (N1) 60 mm, (N2) 120 mm. (N4) Hydrogenation furnace, 1200°C for sulfur, 1000° for fluorine length 120 mm. (O) Tube with joint on stopcock (H). (P) Safety vent on tube (I) (cf. Figure 24). If exit (J) were obstructed, hydrogen could escape through (P) and would not pass into (L), which could cause explosions. (Q) Sidetube with silicone rubber stopper for introduction of sample into quartz spoon (S), which is provided with a magnet (R). (T) Silicone rubber stopper. (U) Capillary, inner diameter 2 mm, length 15 mm.

REAGENTS

Dilute phosphoric acid: Dilute 10 mL of concentrated phosphoric acid to 100 mL with water.

Zinc acetate stock solution: Dissolve 250 g of zinc acetate,

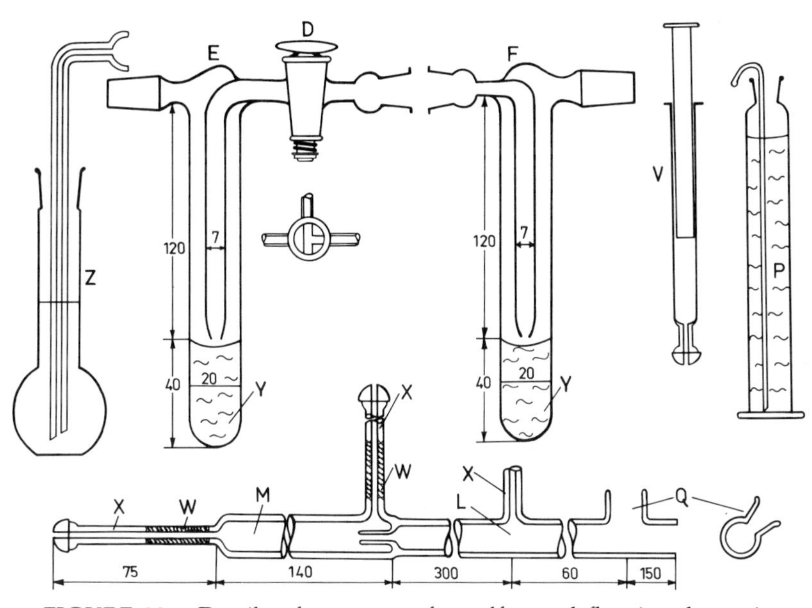

FIGURE 29. Details of apparatus for sulfur and fluorine determination. [Reprinted with permission from Kirsten and Shah (1975). Copyright 1975 American Chemical Society.] (D, E, F) Details of scrubbers (cf. Figure 28). Scrubbers are filled with 40% potassium hydroxide solution below orifices of inlet tubes. Their main purpose is to keep the soda lime in tubes (G) moist and active. Potassium hydroxide is used to lower the water vapor pressure. Evaporated water must be replaced now and then. (Z) Volumetric absorption flask with inlet tube of glass for sulfur, inlet tube of quartz and absorption flask of polypropylene for fluorine. (L, M) Combustion-hydrogenation tube (cf. Figure 28). (P) Safety vent, water-filled (cf. Figure 28). inner diameter of inlet tube, 1 mm; length, 350 mm. (V) Glass syringe with ball joint. (W) Short lengths, 30 mm of opaque quartz capillary sealed in place to avoid strong heat radiation inside clear quartz. The radiation would cause polymerization of silicone grease and nontight joints. (X and W) Quartz capillary inner diameter 2 mm, outer diameter 8 mm. In fluorine determination it is important that there is no cavity inside the ball joint because fluorine would be retained there, which would necessitate a longer sweeping time in the analyses. All conical joints are B14 except for that on (Z), which is B10, and all ball and socket joints are 12/2, lubricated with silicone grease. Thick-walled quartz tubing is used throughout because of its much longer lifetime. Dimensions are millimeters.

$Zn(CH_3COO)_2 \cdot 2H_2O$, in water and dilute to 1 liter. Keep solution in a plastic bottle.

Absorption solution: Dilute daily 10 mL of zinc acetate stock solution to 250 mL with water.

DEP reagent: Dissolve 8.0 g of *N,N*-diethyl-*p*-phenylenediamine monohydrochloride in 400 mL of 9 *M* sulfuric acid and dilute under cooling to 2 liters with water. Keep the solution in a dark bottle in the refrigerator.

Fe reagent: Dissolve 72.0 g of fresh ammonium iron (III) sulfate, $(NH_4Fe(SO_4)_2 \cdot 12H_2O$, in water. Filter through a glass filter to remove small amounts of insoluble material and dilute to 2 liters with water. Keep in refrigerator.

Sodium lauryl sulfate–EDTA solution: Dissolve 200 g of sodium lauryl sulfate and 150 g of ethylenediaminetetraacetic acid disodium salt ($C_{10}H_{14}N_2Na_2O_8 \cdot H_2O$) in water and dilute to 5 liters with water.

ADDITIONAL REAGENTS FOR EXTRACTION METHOD

EDTA solution: Dissolve 30 g of ethylenediaminetetraacetic acid disodium salt in water and dilute to 1 liter.

Salicylic acid solution: Dissolve 0.7 g of salicylic acid in 500 mL of water.

Organic solvent: Mix 380 mL of 1,1,2,2-tetrachloroethane with 20 mL of 2-octanol.

ADDITIONAL REAGENT FOR BISMUTH-, IODINE-, MERCURY- or SELENIUM-CONTAINING SAMPLES

Reducing solution: Dissolve 2.5 g of sodium hypophosphite, $NaH_2PO_4 \cdot H_2O$, in 25 mL of glacial acetic acid and 100 mL of hydroiodic acid, specific gravity 1.7, in a 200-mL round flask with a reflux condenser and a gas inlet tube. Boil under reflux for an hour, bubbling nitrogen through the solution at a rate of about 50 mL/min. Allow to cool under nitrogen and store the well-closed flask in the refrigerator.

ASSEMBLY AND ADJUSTMENT OF APPARATUS

Assemble the apparatus as shown in Figure 28. Start the gas flow so that the oxygen passes through the combustion tube and the hydrogen passes out through the side tube of stopcock (D) while stopcock (H) is turned so that the connection to tube (G) is closed. Turn the furnaces on.

When the furnaces have reached their working temperature, turn stopcock (H) so that tube (G) is connected to tube (O) and the connection to the hydrogenation tube is closed. Turn stopcock (D) so that the hydrogen passes through tube (G) and out through tube (O). Adjust the gas flow rates to 25 mL/min of oxygen and 85 mL/min of hydrogen. Turn stopcock (H) so that the hydrogen passes into the hydrogenation tube. The apparatus is now ready for use.

PROCEDURES

Choose absorption flasks and cuvettes for the spectrophotometric measurement according to Table II. The volumes of reagents for the use of 100-mL absorption flasks follow. For other flasks correspondingly larger or smaller volumes are used.

When samples with very different sulfur contents must be analyzed, it is convenient to use the cuvettes for the lowest-occurring contents and to use different sizes of absorption flasks. The results

TABLE II

Suitable Flasks, Cuvettes, and Volumes of Extraction Solvent for Different Amounts of Sulfur in Sample

Sulfur in sample (μg)	Volume of absorption flask (mL)	Volume of organic phase (mL)	Light path of cuvette (cm)
0–0.3	30	1.5	5
0–1.5	30	3.0	2
0–2.0	30	—	20
0–9	30	—	5
0–15	30	—	3
0– 50	100	—	2
0–250	500	—	2

can then be read from the same calibration curve, and they need only be corrected for the different volumes of the flask.

In our laboratory most analyses are carried out with a 30-mL absorption flask and a 20-cm tube cuvette with black glass walls and a volume of 25 mL, especially made for us by Hellma GmbH Müllheim (Baden, FRG) for use with the long-cell compartment of the Zeiss PMQII spectrophotometer. The light beam of the spectrophotometer is masked to prevent it from striking the walls of the cuvette.

Weigh out the sample in a platinum boat and add 5 μL of dilute phosphoric acid, wetting the sample as much as possible. Turn stopcock (H) so that the hydrogen passes out only through tube (O). Draw spoon (S) into position below stopper (Q). Take out stopper (Q) and put the sample into the spoon. Replace the stopper and turn stopcock (H) so that the hydrogen passes into the hydrogenation tube. Charge the absorption flask with 10 mL of zinc acetate and 50 mL of water and attach it to joint (J). If the sample contains less than 250 μg of organic material, push the sample quickly into the furnace (N1) with the magnet. If it contains more, use furnace (N1) as a movable furnace: Push it back from the combustion tube already before the introduction of the sample and move it to the sidetube (K). After placing the wetted sample into the spoon, move the latter to a spot about 3–4 cm from the furnace (N2). Draw furnace (N1) across the tube, and let it move slowly over the sample to furnace (N2).

Five minutes after the completed combustion, detach the absorption flask from joint (J) and add 5.0 mL of DEP reagent through the gas inlet tube with the syringe (V) (Figure 29) and rotate the flask to mix the solutions. Add 5 mL of Fe reagent through the inlet tube with another syringe (V), draw up the inlet tube, stopper the flask, and shake it immediately and violently for 1 min. Add 10 mL of sodium lauryl sulfate–EDTA and mix by gently turning the flask several times. Fill to the mark with water and mix gently again. Do not shake: shaking would produce foam. Measure the light absorbance immediately at 670 nm using water as the reference.

The following extraction method is recommended: Use 30-mL flasks for the absorption and develop the color as described previously. Add 3 mL of EDTA solution instead of the lauryl sulfate–EDTA. Add 2 mL of salicylic acid solution and the organic solvent. Shake strongly for 30 s and cool in ice water for 2 min. Shake again strongly for 30 s, decant most of the water phase, and pour the organic phase into a centrifuge tube. Spin for 30 s. Pipette clear organic phase into the cuvette and measure the light absorbance at 657 nm, using pure organic phase as the reference.

The following procedure for bismuth-, iodine-, mercury-, or selenium-containing samples is recommended: Introduce the sample into the apparatus as described earlier. Introduce 3 mL of reducing solution into limb (AB) of the scrubber (Figure 30) and about 2 mL of water into limb (AC). Cold water passes through the condenser (AD). Turn stopcock (H) so that the hydrogen passes through the hydrogenation chamber and attach the scrubber to tube (J). Wait until the reduction solution refluxes slowly and is completely decolorized. Attach the charged absorption tube and burn the sample. Detach the absorption tube from the scrubber and develop and measure the color.

Detach the scrubber, turn stopcock (H) so that the hydrogen

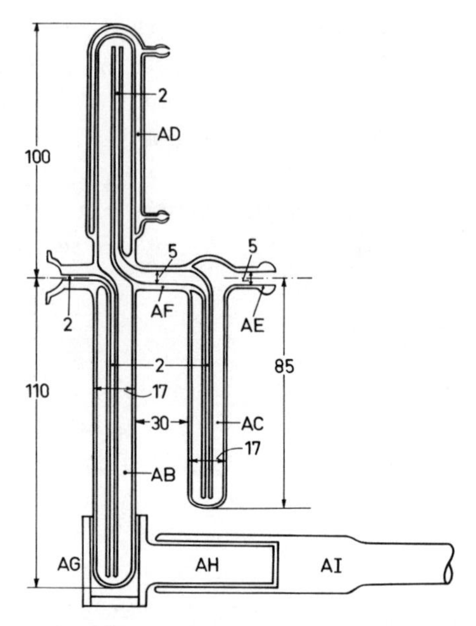

FIGURE 30. Scrubber for elimination of interfering elements in sulfur determination. (AB) Limb of scrubber for reducing solution. (AC) Limb of scrubber for water. (AD) Reflux condenser. (AE) Ball joint. (AF) Connecting tube. (AG) Thick-walled cup of stainless steel, hard-soldered to brass rod (AH), which is held by the soldering iron (AI). The cup, heated by the soldering iron, keeps the reducing solution at boiling temperature. Dimensions are millimeters.

passes out through tube (O) and introduce the next sample. Turn stopcock (H) again to pass the hydrogen into the hydrogenation chamber, reattach the scrubber and absorption tube, and burn the next sample. Repeat this process.

No oxygen should ever be passed through the scrubber because this would oxidize the reducing solution. One scrubber filling is usually sufficient for a day's analyses.

When fractions of micrograms of sulfur are to be determined accurately, carry out a combustion–hydrogenation with an unweighed sample first in the morning in order to equilibrate the apparatus.

DISCUSSION

The determination of sulfur through hydrogenation to hydrogen sulfide is very attractive because hydrogen sulfide can be measured with very high sensitivity and very high specificity: Oxidimetric measurement by oxidation to sulfate gives eight equivalents per sulfur atom, and the ethylene blue method (Rees *et al.*, 1971; Kirsten, 1978b) gives a molar absorbance per atom of sulfur of 7×10^4. It obeys Beer's law.

Several methods for the determination of sulfur through catalytic hydrogenation have been described. They were, however, liable to many interferences: The catalysts were poisoned or covered by carbon and complete recovery of sulfide was not obtained. When the hydrogenation was carried out after a complete combustion, quantitative hydrogenation could be achieved at 1200°C without a catalyst (Kirsten, 1950, 1962). It is very important that the combustion be complete. Incomplete combustion, even without formation of soot, causes low results.

The temperature of the combustion tube should not exceed 700°C. At higher temperatures very small amounts of quartz are volatilized and small blanks are obtained. Surprisingly, the hydrogenation compartment can be kept at much higher temperatures without giving blanks. A layer of silicon probably is formed on the inside of the tube in the hydrogen atmosphere, which prevents the volatilization of quartz. The hydrogenation compartment must be kept at 1200°C to grant a complete hydrogenation.

Sulfur oxides are adsorbed on the quartz walls of the combustion tubes, particularly when these have been attacked by many analyses. Long sweeping times therefore had to be used, and the tubes had to be cleaned frequently. This trouble was avoided completely through the addition of orthophosphoric acid to every sample. The acid expels sulfur from metal salts, and it occupies any alkaline sites on the quartz walls, which could adsorb sulfur (Kirsten, 1967). Furthermore, it forms a very stable layer on the inside of the combustion tube (Binkowski and Gizinsky, 1977). Too much phosphoric acid should not be used. When ten times the recommended amount was used, lower and less reproducible results were obtained. The desorption of sulfur with phosphoric acid is so complete that trace determinations can be carried out right after the analysis of alkali salts with high sulfur content.

Commercial hydrogen sometimes contains traces of sulfur-containing impurities. When furnace N3 is hot, they are decomposed and the sulfur is retained in the soda lime. Before the furnace is hot they pass through it and are partially adsorbed on the soda lime, from which they are slowly desorbed, causing high but slowly decreasing blanks. This is avoided by letting the hydrogen escape through stopcock (D) until the furnace is hot.

No interferences occurred when the following substances were added to samples of sulfur-containing compounds: 4 mg of ammonium fluoride, NH_4F; 5 mg of antimony, Sb, metal; 3 mg of arsenic oxide, As_2O_3; 7 mg of cadmium acetate, $Cd(C_2H_3O_2)\cdot 3H_2O$; 8 mg of cadmium, Cd, metal; 4 mg of copper, Cu, metal: 25 mg of gallium, Ga, metal; 12 mg of germanium, Ge, metal; 4 mg of iron chloride, $FeCl_3\cdot 6H_2O$; 4 mg of lead, Pb, metal; 10 mg of silver nitrate, $AgNO_3$; 3 mg of tin chloride, $SnCl_2\cdot 2H_2O$; 5 mg of tungstic acid, H_2WO_4; 5 mg of vanadium oxide, V_2O_5.

The following compounds interfere unless the scrubber with reducing solution is used:

1 mg of iodine, I_2: no color develops.

4 mg of mercury(I) nitrate, $Hg_2(NO_3)_2\cdot 2H_2O$: about 30% low results.

8 mg of bismuth(III) nitrate, $Bi(NO_3)_3\cdot 5H_2O$: no color develops.

The hydrogen sulfide formed in the combustion–hydrogenation method can also be measured oxidimetrically by oxidation to sulfate with hot alkaline iodate solution and back-titration of the excess of iodate with thiosulfate (Kirsten, 1959). This method is not as sensitive as the spectrophotometric method, but it is very accurate, and iodine does not interfere.

CHAPTER 16

ULTRAMICRO DETERMINATION OF FLUORINE*

The same apparatus as that described for the determination of sulfur is used. Gas inlet tubes and absorption flasks [(Z) in Figure 29] are made of quartz or plastic material.

The sample, covered with tungsten(VI) oxide and wetted with orthophosphoric acid or wetted only with a phosphate flux, is burned in a flow of oxygen. The combustion gases pass into a hot chamber containing an excess of hydrogen. The fluoride is absorbed in water and determined by spectrophotometry after reaction with lanthane alizarin complexan.

Aluminum, iron, and other metals, which interfere with many determination methods, remain in the combustion boat, and sulfate and phosphate, which also interfere, are reduced to sulfide and phosphine.

*See Kirsten and Shah (1975) and Belcher (1966).

REAGENTS

Dilute phosphoric acid: Dilute 10 mL of concentrated phosphoric acid to 100 mL with water.

Phosphate flux: Dissolve 14 g of sodium dihydrogen phosphate, $NaH_2PO_4 \cdot H_2O$, in the least possible amount of water, add 3.5 mL of concentrated phosphoric acid, and dilute to 25 mL with water.

Tungsten(VI) oxide: Heat tungsten(VI) oxide powder or tungstic acid to 900°C in a flow of oxygen for about 1 hour. Cool and keep in a well-closed plastic bottle.

Potassium hydroxide: Dissolve 100 g of analytical grade potassium hydroxide, 85%, in 100 mL of water.

Sodium hydroxide, 1 *M*.

Hydrochloric acid, 1 *M*.

Alizarin complexan solution: Weigh out 241 mg of alizarin-3-methylamine-*N*,*N*-diacetic acid dihydrate, Merck (Darmstadt, FRG) into a 250-mL volumetric flask, and dissolve it in the least possible amount of freshly prepared 1 *M* sodium hydroxide. Dilute to 50 mL with water and check that all is really dissolved and that the solution has a blue-violet color. If not, add more sodium hydroxide. Add 125 mg of sodium acetate, $NaAc \cdot 3H_2O$. When all is dissolved, slowly add 1 *M* hydrochloric acid until the solution just becomes red-violet, pH between 5 and 6. Add 25 mL of acetone and make up to the mark with water.

Lanthanum nitrate: Dissolve 271 mg of lanthanum nitrate, $La(NO_3)_3 \cdot 6H_2O$, in water and dilute to 250 mL.

Acetate buffer: Dissolve 105 g of sodium acetate, $NaAc \cdot 3H_2O$, in 100 mL of glacial acetic acid in a beaker with careful heating. Dilute with water to 500 mL. Cool to room temperature and make up to 500 mL.

Sodium acetate solution: Dissolve 25 g of $NaAc \cdot 3H_2O$ in water and dilute to 500 mL.

Color reagent: Mix 50 mL of Alizarin complexan solution, 50 mL of acetate buffer, 50 mL of lanthanum nitrate, and 250 mL of acetone. The solution is stable for at least 2 weeks.

Prepare and keep all reagents in well-closed plastic vessels.

ASSEMBLY AND ADJUSTMENT OF APPARATUS

Assemble and adjust the apparatus as described under determination of sulfur. Combustion furnaces may have higher temperature, 800–1000°C, and reduction furnaces lower temperature, about 1000°C. The gas flow rates should be 25 mL/min for oxygen and 80 mL/min for hydrogen.

PROCEDURES

Weigh out the sample in a platinum boat. Add 5 μL of dilute phosphoric acid to the sample, taking care to wet the sample as well as possible and introduce the boat into the spoon through opening(Q). If the sample is inorganic or contains ashes, mix it well with an equal volume of tungsten(VI) oxide before adding the phosphoric acid to the boat. Some inorganic minerals like calcium phosphate will not liberate their fluorine content with tungsten(VI) oxide. They are, however, decomposed and liberate the fluorine when they are wetted with 5 μL of phosphate flux. No tungsten oxide or orthophosphoric acid is used.

Close stopper (Q). Turn stopcock (H) so that the hydrogen passes into chamber (M). Attach the absorption vessel—a 20 mL polypropylene measuring cylinder with 2 mL of water, or a larger absorption flask (Z). If the sample contains less than 250 μg of organic material, push it quickly into the furnace with the magnet. If it contains more, use furnace (N1) as a movable furnace: Push it back from the combustion tube before introducing the sample and move it to the side tube (K). After placing the sample into the spoon move the latter to a spot about 3–4 cm from furnace (N2).

Draw furnace (N1) across the tube and let it move slowly over the sample to furnace (N2).

Ten minutes after the completed combustion detach the absorption vessel from joint (J). If the 20-mL measuring cylinder was used, wash the inlet tube with a few drops of water. Add 6.000 mL of color reagent and dilute to the mark with water. Let stand in the dark for at least 90 min and read at 620 nm using the same reagent mixture in the reference cuvette.

One μg of fluorine gives an absorbance of 0.315 in 10-cm cuvettes. The calibration curve is a straight line. With shorter cuvettes up to 12 μg can be measured. Above that the calibration curve is bent.

Larger amounts of fluorine can be determined by using correspondingly larger absorption flasks and correspondingly larger volumes of color reagent—or better, by absorbing the fluoride in a 50-mL flask, making up to the mark with water, and pipetting an aliquot of the solution into another volumetric flask for the color development. Plastic pipettes and flasks must be used.

CHAPTER 17

SIMULTANEOUS DETERMINATION OF SULFUR AND FLUORINE*

The sample undergoes combustion and is hydrogenated as described for the determination of sulfur. Fluoride is absorbed in a flask containing a few milliliters of water. Hydrogen sulfide passes through this solution and is absorbed in a following flask with zinc acetate. They are measured separately by spectrophotometry.

APPARATUS

The apparatus shown in Figures 28 and 29 is used together with the absorption train shown in Figure 31.

*See Kirsten and Shah (1975).

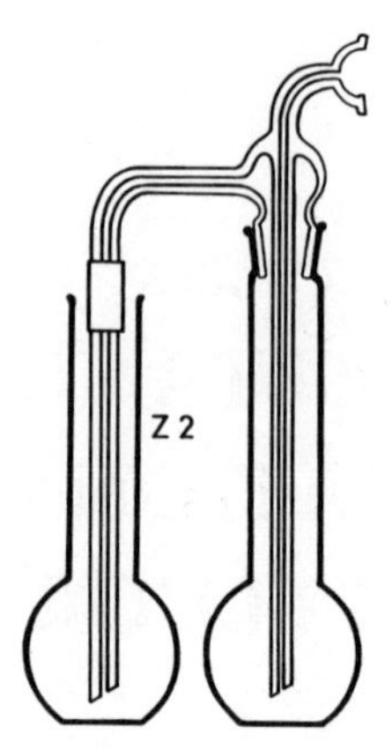

FIGURE 31. Absorption train for simultaneous determination of sulfur and fluorine. [Reprinted with permission from Kirsten and Shah (1975). Copyright 1975 American Chemical Society.] (Z2) Volumetric absorption flasks of polypropylene. The inlet tube has ground joints of quartz. The width of the capillaries is 2 mm.

PROCEDURE

Adjust the temperatures of the furnaces and carry out the combustion-hydrogenation as described for the determination of sulfur. Charge the first flask of the absorption train with a few milliliters of water and the second with zinc acetate and water. Allow 10 min of sweeping time after the combustion-hydrogenation.

After the absorption, first detach the second flask and develop and measure the ethylene blue color; then detach the first flask and develop and measure the fluorine color.

CHAPTER 18

TRACE DETERMINATION OF SULFUR OR FLUORINE*

APPARATUS

Use the apparatus shown in Figures 32 and 33.

ADDITIONAL REAGENTS FOR FLUORINE (SELDOM NEEDED)

Nitrophenol: Dissolve 200 mg of *P*-nitrophenol in 100 mL of water.

Absorbent: Mix 5 mL of nitrophenol solution with 95 mL of sodium acetate solution.

*See Kirsten (1976a, 1979b).

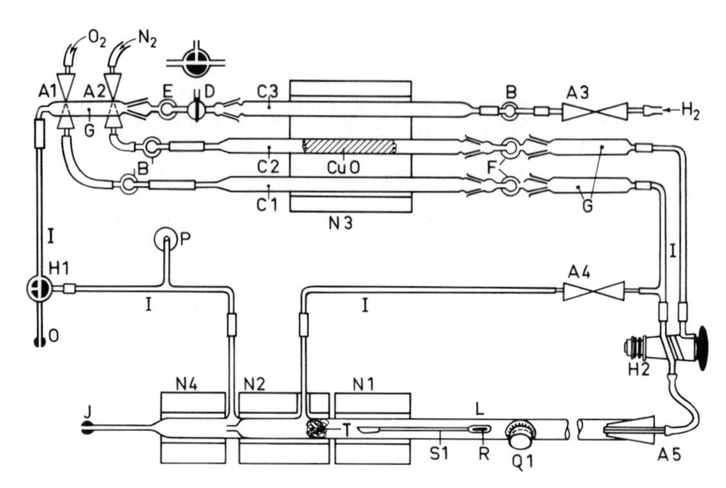

FIGURE 32. Assembly for trace determination of sulfur and fluorine. [Reprinted with permission from Kirsten (1976a). Copyright 1976 American Chemical Society.] (A1–A4) Variable restricting valves. (A5) Restricting capillary. (B) Rotameters. (C) Quartz tubes in furnace (N3). (D) Three-way stopcock on scrubber (E) (cf. Figure 29). (F) Scrubber (cf. Figure 29). (G) Tubes filled with moist soda lime. (H1) Three-way stopcock. (H2) Two-way stopcock. (I) Glass capillary connections. (J) Ball joint for absorption flask. (L) Combustion–hydrogenation tube, quartz, inner diameter 12 mm, outer 15 mm. (N1, N2) Split-type furnaces. (N3) Tube furnace with three holes. (N4) High-temperature tube furnace. Lengths of furnaces 120 mm. Temperatures for sulfur: (N1, N2) 700°C, (N3) 850°C, (N4) 1200°C; temperatures for fluorine: (N1,N2) 900°C, (N3) 850°C, (N4) 1050°C. Furnace (N1) can be made to travel along the compustion tube as usual in elemental analysis. (O) Tube with ball joint on stopcock (H1). (P) Safety vent on tube (I), (cf. Figure 29). (Q1) Sidetube with silicone rubber stopper for introduction of sample into spoon (S1), which is provided with magnet (R). (T) Wad of quartz wool that prevents tungsten(VI) oxide from getting into the second combustion chamber. It is *only used for fluorine* determination.

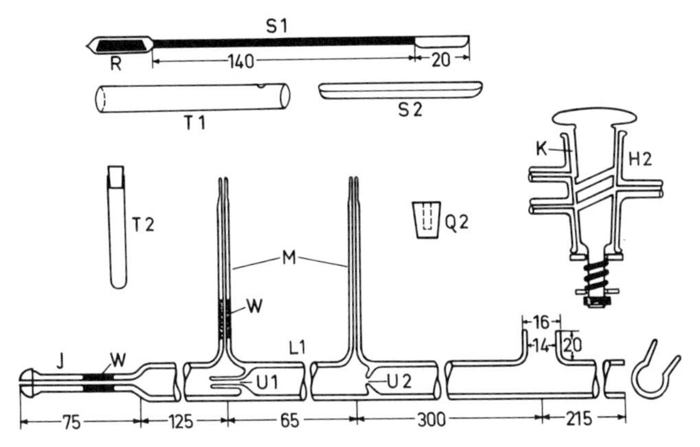

FIGURE 33. Details of assembly for trace determination of sulfur and fluorine. [Reprinted from Kirsten (1976a). Copyright 1976 American Chemical Society.] (H2) Stopcock with groove (K), which lets nitrogen escape when oxygen is passed into the combustion tube through capillary (A5) (Figure 32). (J) Capillary tube, inner diameter 3 mm with ball joint. (L1) Second combustion chamber. (U1) Capillary, inner diameter 3 mm. (U2) Orifice, 4 mm. (M) Sidetubes for introduction of oxygen or hydrogen. (Q2) Silicone rubber stopper with hole for introduction of liquids into the combustion tube through opening (Q1) (Figure 32), with a syringe. (S1) Spoon of quartz for samples with magnet (R). (S2) Quartz boat for sample mixed with tungsten(VI) oxide. (T1) Protection tube of quartz for boat (S2). (T2) Glass tube with stopper of silicone rubber for mixing of sample with tungsten(VI) oxide. (W) Opaque pieces of tubing that prevent heat radiation inside quartz tubes. All rubber connections of the apparatus are made with silicone rubber tubing. Other details of the apparatus are the same as those shown in Figures 28 and 29. Dimensions are millimeters.

ADJUSTMENT OF APPARATUS

Set up the apparatus as described earlier for the ultramicro determination. Adjust the gas flow rates to 75 mL of oxygen, 20 mL of nitrogen, and 250 mL of hydrogen per minute. Adjust needle valve (A4) and capillary (A5) so that only a few milliliters per minute of oxygen pass through (A4) when stopcock (H2) is opened but so that the total flow of oxygen does not decrease very much when (H2) is closed. Adjust the temperatures of the furnaces as described earlier under ultramicro determination of sulfur or fluorine.

COMBUSTION METHODS

Solid Inorganic Samples and Samples with Very Low Contents of Organic Material

Push back furnace (N1) from the combustion tube and remove spoon (S1). Weigh the finely ground sample into a small, glass-stoppered specimen tube (T2). Add an equal volume of tungsten(VI) oxide. If the sample is alkaline—such as lime, magnesia, sodium carbonate, or hydroxides—add at least twice the weight of the tungsten(VI) oxide that is necessary to neutralize the sample. Mix thoroughly and pour the mixture into a big quartz boat (S2). If the sample is small or adheres to the walls of the specimen tube, wash the latter with a few small portions of tungsten(VI) oxide and add the washings to the boat. Add 5 μL of dilute phosphoric acid to the upstream end of the boat. Introduce the boat into the protection tube (T1) and introduce protection tube and boat into the combustion tube through the opening at (A5) so that the boat is at a distance of about 3 cm from furnace (N2). Close (A5) and turn stopcock (H2) so that oxygen passes into the combustion tube through (A5) and turn stopcock (H1) so that hydrogen passes into the hydrogenation chamber. Attach the absorption vessel and draw furnace (N1) across the tube with the sample. After combustion is completed, detach the absorption tube and turn stopcock (H1) so that the hydrogen passes out through joint (O).

Some kinds of samples (e.g., calcium phosphate and other refractory minerals) are not decomposed by tungsten(VI) oxide, and their fluorine is not liberated. In this case add phosphate flux to wet the sample instead of tungsten(VI) oxide and orthophosphoric acid.

Samples that Contain Large Amounts of Elements Whose Decomposition Products Can Change the pH of the Absorption Solution (Hardly Ever Needed)

Absorb in 1 mL of absorbent. If the solution turns yellow, add 1 *M* hydrochloric acid drop by drop until the solution just becomes colorless. Agitate with the inlet tube and lift up the inlet tube so that the solution inside it also is neutralized. If the solution is colorless, add first, drop by drop, 1 *M* sodium hydroxide until the solution is yellow; then add hydrochloric acid, as described earlier. Then add color reagent or make up the solution to the mark with water and take an aliquot for the color development and measurement.

Samples Containing Large Amounts of Organic Material

Prepare the samples and introduce them into the apparatus as described earlier for inorganic substances. Close (A5) and turn stopcock (H2) so that nitrogen passes into the combustion tube, and then (H1) so that hydrogen passes into the hydrogenation chamber. Attach the absorption vessel and draw furnace (N1) across the tube somewhat upstream from the boat, and let it slowly move to furnace (N2). When no more gases are developed from the sample, turn stopcock (H2) so that oxygen passes over the boat and the sample is completely burned. Detach the absorption vessel and turn stopcock (H1) so that the hydrogen passes out through joint (O).

Oils and Liquids that Can Give Explosive Vapors

Use spoon (S1) and stopper (Q2). Push back furnace (N1) from the tube. Turn stopcock (H2) so that nitrogen passes into the combustion tube, and, with the magnet, draw the spoon to the opening for stopper (Q2). Open the stopper and introduce 5 μL of dilute phosphoric acid into the spoon. Stopper again, turn stopcock (H1) so that hydrogen passes into the hydrogenation chamber, and charge the absorption vessel and attach it to joint (J). Introduce the sample into the spoon through stopper (Q2) with a syringe. Move the spoon to a position near furnace (N2), and let the sample volatilize slowly. When nothing more is evaporated, draw furnace (N1) over the tube where the rod of spoon (S2) is situated and let the furnace move slowly over the sample. When it has reached furnace (N2), turn stopcock (H2) so that the oxygen passes through the tube and burns away the residues of the sample. When all is burned, allow another 5–10 min of sweeping. Detach the absorption flask and develop and measure the color as described for ultramicro determination.

No tungsten(VI) oxide powder should be allowed to get into the second combustion chamber (L). The color of the fluorine reagent is bleached when tungsten (VI) oxide is in (L). When the downstream parts of the tube are contaminated with tungsten(VI) oxide, take out the tube and treat it with warm sodium hydroxide solution until it is colorless. Wash it well with warm water and allow it to stand filled with diluted nitric acid 1 : 20 for an hour. Wash again with water and alcohol and dry.

When sulfur is determined, it is very important to carry out the combustion of large samples slowly. Incomplete combustion causes low sulfur results. In the fluorine determination correct results are obtained even when small amounts of soot pass through the tube. No quartz wool (T) is used when sulfur is determined.

CHAPTER 19

ULTRAMICRO DETERMINATION OF CHLORINE, BROMINE, AND IODINE*

In the hot-flask method the sample, weighed out in a small platinum boat, is burned by static flash combustion in the hot part of a closed, oxygen-filled quartz vessel, as shown in Figure 34. Metal halides are decomposed by orthophosphoric acid, which is added to all samples before the combustion. The combustion gases diffuse from the combustion chamber into the cool part of the vessel below the furnace, where they are absorbed in the small ground joint cup in an organic buffer solution.

The cup with the absorbing solution is then removed, and the halides are titrated in it using an automatic argentometric dead-

*See Kirsten (1963), Kirsten *et al.* (1967), and Kirsten (1976b).

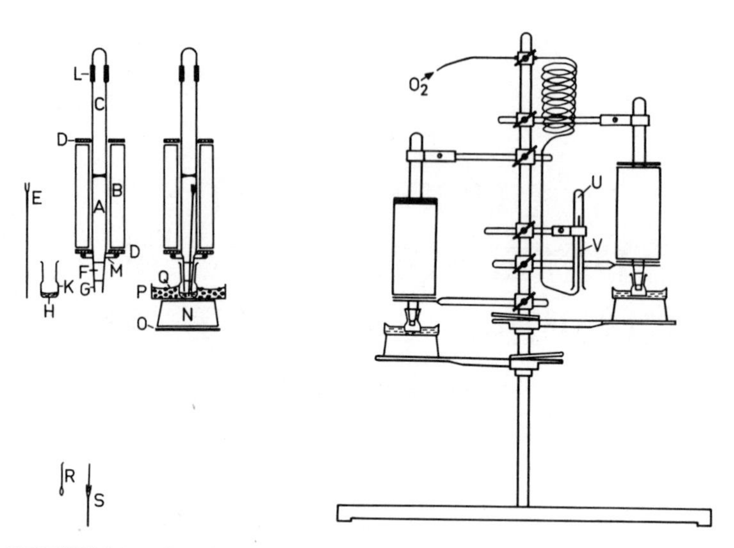

FIGURE 34. Combustion apparatus for the determination of chlorine, bromine, and iodine. [From Kirsten, (1976).] (A) Combustion vessel of quartz, inner diameter about 14 mm, wall thickness about 1.5 mm, length of part in furnace about 60 mm. (B) Tube furnace, temperature 850–900°C. (C) Handle of opaque quartz tubing sealed onto combustion vessel (A). (D) Shield of insulating material. (E) Sample holder: A quartz cup, inner diameter 3 mm, is sealed on a quartz rod, diameter about 2 mm. Length of cup is about 6 mm, total length of sample holder is about 95 mm. (F) Joint B12 with extension (G), which ends about 2 mm above absorption solution (H) when the sample has been inserted. (K) Vessel with B12 joint fitting to (A). Length of part below joint about 18 mm. (L) Support clamp that holds combustion vessel (A) with handle (C). (M) Ring support that holds (D). (N) Soft rubber stopper securing steady pressure against vessels (P) and (K) from below. (O) Adjustable support table. (P) Metal or plastic vessel. (Q) Ice and ice water. (R) Sample of volatile liquid weighed out in capillary of Supremax glass. (S) Sample holder (E) with capillary (R) inserted. (U) Test tube clamped upside down to protect capillary (V). (V) Stainless steel capillary through which oxygen is blown into the combustion vessels (A). Its straight, upward-pointing end can be inserted into the combustion vessels or into test tube (U). In both cases it is held in place by the tension of its coil.

stop titration method, according to Bishop and Dhaneshwar (1962), as shown in Figures 35 and 36.

The method involves a minimum number of operations: There are no wrappings of samples before the combustion and no transfers of the absorption solution. There is no boiling and no evaporation. The volume of the titration solution is very small, which provides for a sharp endpoint. Both organic and inorganic compounds can be analyzed. Mercury in the samples interferes with the procedure because of its volatility.

It is most convenient to have enough rods (E) (Figure 34), platinum boats, and vessels for one day's work and to clean them at the end of the day.

HOT-FLASK COMBUSTION

Reagents for Combustion

Orthophosphoric acid, analytical grade, specific gravity 1.7.

Tris, tris(hydroxymethyl)aminomethane, analytical grade, 0.3 *M* aqueous solution.

Hydrazinium hydroxide, 99.5%, Merck, for synthesis.

Absorbent: Add 0.20 mL of hydrazinium hydroxide to 60 mL of tris solution. Prepare new absorbent every week.

Adjustment of Apparatus for Combustion

Assemble the apparatus as shown in Figure 34. Heat joint (F) and its extension (G) to red heat with a gas flame before every series of analyses. Adjust the oxygen flow from the steel capillary to about 20 mL/min.

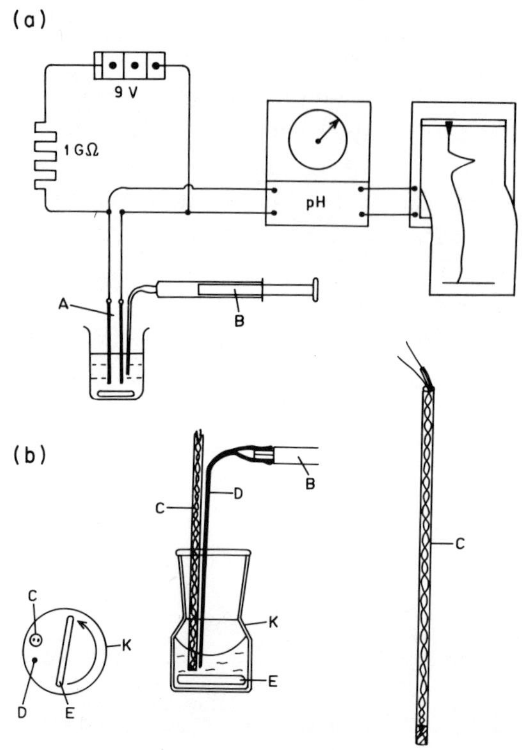

FIGURE 35. Titration apparatus for chloride, bromide, and iodide. [From Kirsten (1976).] (a) Schematic layout of arrangement. (A) Silver electrodes. (B) Motor-driven syringe burette. (b) Arrangement of electrodes in titration vessel (K). The stirrer should move rapidly tightly under the electrodes in such a direction that the added silver nitrate from the burette moves around in the vessel first before it reaches the electrodes. (C) Electrodes baked into polyethylene: Use two wires of pure silver, 100–200 mm long, diameter 0.6 mm. Introduce one of them into a snugly fitting polyethylene tube. Twist the two wires around each other and introduce them into a snugly fitting wider polyethylene tube. Let a few drops of liquid polyethylene fall on the lower end of the tubes to form a tight seal. After cooling, shape the end with a razor blade and a fine, sharp file, so that the clean, flat ends of the silver wire lie in the same plane as the surrounding polyethylene. There should be no fissures between the metal and the plastic because entering liquid would vitiate the measured

Procedure for Combustion

Insert capillary (V) into the first hot-flask. Weigh the sample, containing not more than 500–700 μg of organic material, into a platinum boat. Add 1 μL of phosphoric acid, taking care to wet the sample as much as possible. Place the boat in holder (E). Use a pipe cleaner to lubricate the upper part of the joint of vessel (K) with a very thin layer of Vaseline. Pipette 0.3 mL of absorbent into the vessel taking care not to wet the joint. Place rod (E) in vessel (K). Take capillary (V) out from the hot flask and place it into the other hot flask. Insert rod (E) quickly into the first hot flask and tighten the joint immediately. Hold vessel (K) firmly at the joint with one hand and raise table (O) with stopper (N) and vessel (P) with the other hand and press them firmly against vessel (K). Fill vessel (P) with ice and water (Q). Allow to stand for 20 min. In the meantime weigh out the next sample, replace capillary (V) in testtube (U), and introduce the next sample into the second hot flask. After 20 min, lower table (O) and remove vessel (K) with rod (E). Insert capillary (V) into the hot flask. Start the titration.

Weigh out volatile liquids in Supremax capillaries (R) (Figure 34). Fill and close them in the usual manner and place them in holder (E) as shown in inset (S). The thin end of the capillary is crushed when it meets the upper wall of the hot flask during the insertion. Check the temperature of the furnace before using this method. If it is much higher than 850°C, splinters of Supremax glass may be fused to the walls of the quartz flask.

---

potentials. (D) Burette tip of polyethylene mounted on the syringe burette (B). (E) Magnetic stirrer made of about 0.6-mm-thick iron wire sealed into snugly fitting thin-walled glass or polyethylene tubing.

The 9-V battery (two 4.5-V torch batteries) is mounted into a grounded metal case together with the resistor 1 G Ω. The batteries have a lifetime of about 3 years. The burette should add silver nitrate at a constant rate, 20 μL/min is convenient. The sensitivity of the recorder depends on the output of the pH meter. With a Metrohm meter E 300 B a sensitivity of 50 mV is suitable for all three halogens. Chart speed about 2–3 cm/min.

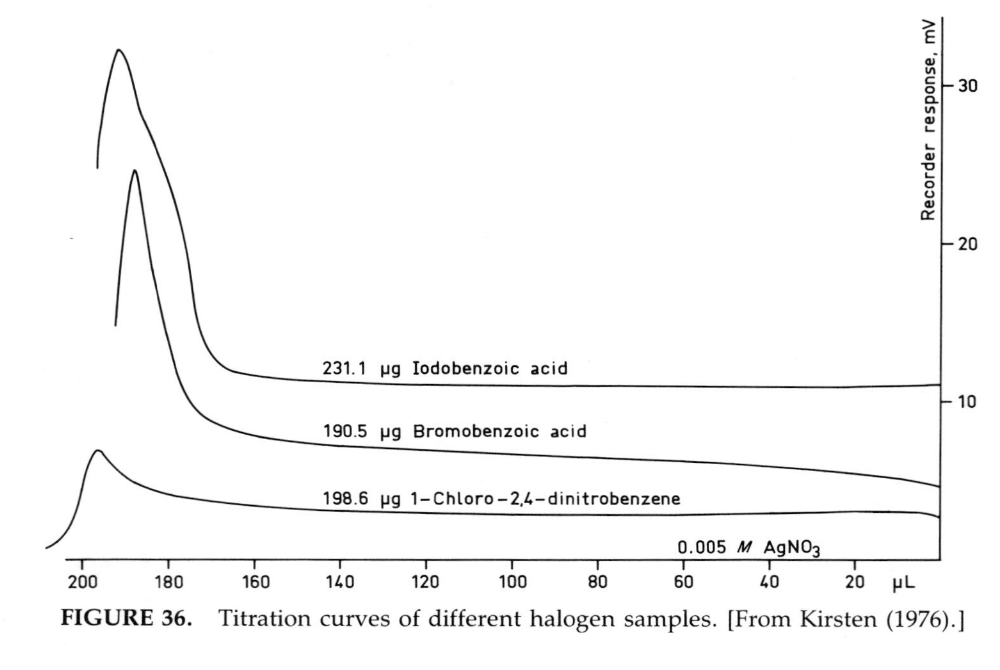

FIGURE 36. Titration curves of different halogen samples. [From Kirsten (1976).]

At the end of the day brush out vessels (K) with water, using a nylon-bristled test tube brush, immerse them, together with the rods and boats, in concentrated sulfuric acid, and heat them to near boiling. Pour off the sulfuric acid, which can be used repeatedly. Let cool and rinse well with distilled water only. Dry in an oven and place, still hot, in a desiccator over soda lime. Keep there until needed. Heat the platinum boats to glowing in a gas flame before use.

If necessary, vessels (K) can be brushed out, immersed in cold sulfuric acid, rinsed with redistilled water, and reused immediately. Do not use tapwater for rinsing.

Turn off the gas flow overnight and close the hot flasks with empty vessels (K).

TITRATION

Adjustment of Apparatus for Titration

Assemble the titration apparatus as shown in Figure 35.

Reagents for Titration

Glacial acetic acid, analytical grade.

Silver nitrate stock solution, 0.1 *M*, accurately prepared. Prepare the titration solution, 0.005 *M*, every month by diluting the stock solution with water. Use 0.002 *M* silver nitrate solution for the analysis of samples below 100 μg.

Procedure

After the hot-flask combustion wash the lower end of rod (E) with 1 mL of glacial acetic acid. Let the acid flow into vessel (K).

Immediately add a magnetic stirring bar, introduce the electrode and burette tip, turn on the stirrer, and start the titrator.

PRECAUTIONS

The laboratory atmosphere is often contaminated with chlorine compounds. Therefore prepare large volumes of stock reagent solutions—1000 mL—and keep them in well-stoppered bottles to decrease the influence of atmospheric contamination. Keep the mixed absorbent and the acetic acid in flasks with Grunbaum pipettes as shown in Figure 1, or in another dispenser, in which they are protected from atmospheric contamination. Start the titration *immediately* after the combustion.

CHAPTER 20

TRACE DETERMINATION OF CHLORINE, BROMINE, AND IODINE*

APPARATUS

The Schöniger flask combustion equipment shown in Figure 37 is used for the combustion of the sample, and the same equipment as described earlier is used for the titration.

REAGENTS

Absorbent for chlorine: Dissolve 30 g of sodium acetate trihydrate in 70 mL of water and add 100 mL of glacial acetic acid.

*See Kirsten (1976b).

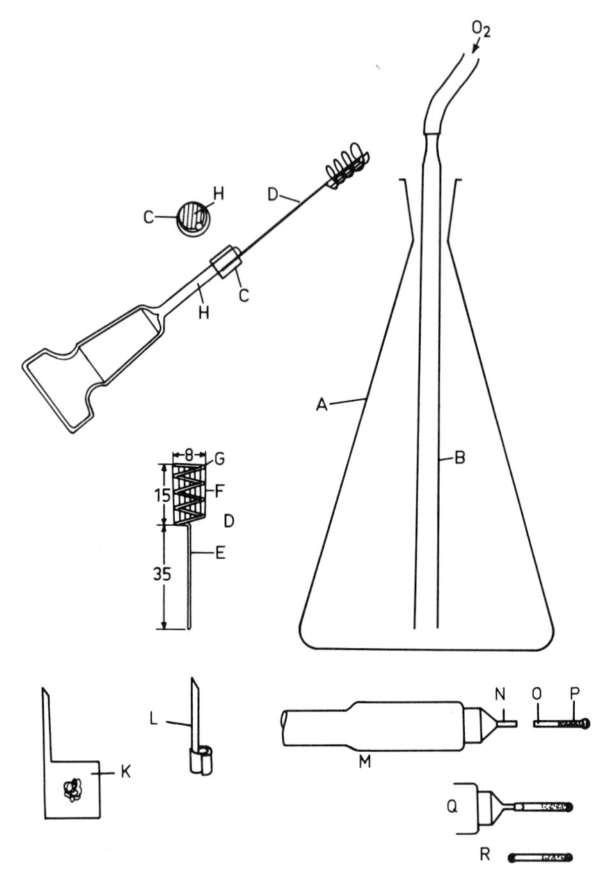

FIGURE 37. Schöniger-flask combustion [From Kirsten (1976b).] (A) 300-mL Erlenmeyer flask with B14 joint. (B) Gas inlet tube. (C) Short piece of polyethylene tubing. (D) Holder of platinum wire. Diameter of stem wire (E) and of spiral wire (G) 1.0 mm, of longitudinal wire (F) 0.7 mm. Holder (D) is held by tubing ring (C) in a short groove ground into the end of glass rod (H) with a dentist's drill. (K) Filter paper with sample. (L) Folded filter paper with sample. (M) Soldering iron with brass rod (N), diameter 1.0 mm coated with silicone grease. (O) Surgical polyethylene tubing, inner diameter 1.14 mm, outer 1.57 mm. (P) Cotton wool and sample, which has been introduced with a syringe or a pipette. (Q) Heating the open end of the sample tube: The end is immediately sealed by

Absorbent for bromine and iodine: Add 25 μL of concentrated hydrazinium hydrate to 10 mL of the chlorine absorbent.

Filter paper: Test for blanks. Thin, ordinary paper often gives lower blanks than ash-free acid-washed paper. Use thin-walled polyethylene tubing or other containers for combustion of volatile compounds.

PROCEDURE

Place the combustion flasks in a line on the table. Blow oxygen into the first one to fill it. Add immediately 1.00 mL of absorbent with a rapid, precise dispenser, and stopper the flask immediately with an ordinary glass or polyethylene stopper. Charge all flasks in the same manner.

Place the first flask into the polyethylene bottom (Figure 38) and put the wire gauze cone over it.

Wrap the sample into a filter paper as shown in Figure 37 and put it into the basket of the combustion stopper with the fuse pointing out. Grasp the bottom of the flask and the cone with one hand and remove the ordinary stopper. Take the combustion stopper in the other hand, ignite the top of the fuse with a microburner, hold the flask horizontally, and immediately insert the stopper with the burning fuse. Hold the flask with the stopper downward

---

pressing it with a forceps or a smooth pair of tongs. (R) Sealed tube, ready for combustion (Kirsten, 1959).

The soldering iron (M) is fed with a variable transformer. The rod (N) is lubricated with silicone stopcock grease, the excess of which is carefully wiped off. The temperature is adjusted so that the end of polyethylene tubing, which is pushed over rod (N), softens, but does not melt. The silicone-treated surface rod (N) is stable for a long time. Dimensions are millimeters.

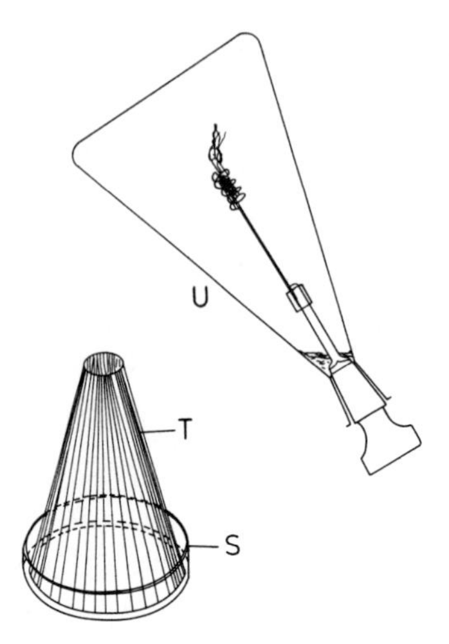

FIGURE 38. Flask combustion and safety jacket. Compare Gouverneur and Eerbeek (1962). Cut off the bottom part (S) of a polyethylene beaker, which fits over the combustion flask, so that about 1.5 cm of the wall remains. Fold or solder a cone (T) of metal wire gauze, which fits over the flask and into the beaker (S) as shown. (U) Combustion.

as shown in inset (U) (Figure 38), firmly holding the stopper in the joint. When the combustion is finished, shake the flask violently, remove the safety jacket, and place the flask on the table for an hour. Put the safety jacket on the next flask and burn the next sample in the same manner. Repeat the process.

Suck high-boiling liquids (e.g., oils) into a micropipette and weigh it. Pipette the liquid onto a ready, folded filter paper in a platinum holder and burn it immediately. Reweigh the pipette.

Weigh out volatile liquids in polyethylene tubing as shown in Figure 37. Close a short tube of polyethylene tubing at one end by heating and pressing the end with a flat, smooth forceps. Press a small amount of cotton into the tube, taking care that nothing protrudes at the open end. Weigh the tube and introduce the sample to the bottom of the tube with a syringe. Push the open end of the tube over the rod (N), so that about 2–3 mm of the rod are in the tube, which is heated to soften. Draw back the tube and press its end together immediately with a forceps or a pair of pliers. Weigh the tube, wrap it into a small filter paper, and burn it in the same manner as the other samples.

When the first flask has been standing for about an hour, loosen and lift the stopper slightly and press on the polyethylene ring with a spatula, so that ring and platinum holder fall into the flask. Wash the stopper and the joint with 2.00 mL of glacial acetic acid. Fix the stopper again and shake vigorously so that the solution is mixed well and any residues on the platinum are washed off.

Pipette 2.00 mL of the solution into the titration vessel and titrate with 0.005 *M* silver nitrate.

CHAPTER 21

ULTRAMICRO AND TRACE DETERMINATION OF PHOSPHORUS AND PHOSPHATE*

The apparatus shown in Figure 39 is used together with a centrifuge, which can accommodate the digestion flasks with the stoppers on, an ultraviolet spectrophotometer, and a test-tube shaker (Figure 40).

REAGENTS

Digestion solution, ultramicro: Mix 100 mL of nitric acid, 16.0 *M*, with 10 mL of perchloric acid, 11.6 *M*, add 95 mL of sulfuric acid, 18

*See Kirsten and Carlsson (1960), and Kirsten (1967).

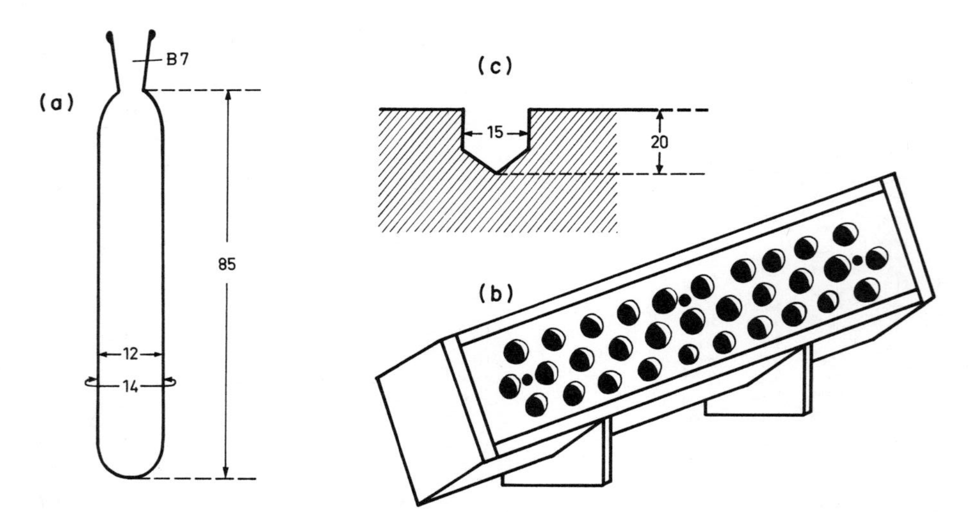

FIGURE 39. Digestion equipment for phosphorus determination. (a) Digestion flask of borosilicate glass with ground joint B7. (b) Heating block of stainless steel or aluminum with holes (c) for digestion flasks and with three holes for thermometers. For ultramicro analysis the block is kept constantly at 250°C. For trace analysis it is connected to a time switch, which provides it with different adjustable voltages for different adjustable periods from a stepped transformer. A convenient apparatus for programmed heating, designed by Knapp (1975), is available from Hans Kürner, Analysentechnik (Neuberg, FRG). Dimensions are millimeters.

FIGURE 40. Vibration shaker Vibrax VXR with holder VX2 for 36 flasks, from Janke and Kunkel, IKA-Werk, D-7813 (Staufen, FRG) for extraction of molybdophosphoric acid with amyl acetate.

M, and, after cooling to room temperature, dilute to 250 mL with nitric acid, 16.0 *M*.

Digestion solution, trace: Mix 200 mL of nitric acid, 16.0 *M* with 150 mL of perchloric acid, 11.6 *M*, add 95 mL of sulfuric acid, 18.0 *M*, and, after cooling to room temperature, dilute to 500 mL with nitric acid, 16.0 *M*.

Molybdate solution: Dissolve 120.0 g of sodium molybdate, $NaMoO_4 \cdot 2H_2O$ in about 600 mL of water and add 2 mL of sulfuric acid, 18.0 *M*. Extract three times with 100 mL of 1-butanol every time. Suck away all butanol with a pipette and a filter pump every time. Extract twice with 100 mL of ethyl ether every time and then extract twice with 100 mL of hexane. Suck away all hexane. Dilute the water phase to 1000 mL with water.

Use chemicals of highest available purity for these reagents.

Amyl acetate: Amyl acetate of highest spectral purity is not commercially available. Merck 1231 isoamyl acetate is sufficiently pure for measurement of reasonably small amounts of phosphorus in 1-cm cuvettes. It can be purified satisfactorily for measurement in 5-cm cuvettes by column distillation. Hopkin and Williams ANALAR 1600 amyl acetate has a lower spectral purity, but it can be purified to a high degree in the following manner: Fill a 40-mm-wide, 1-m-long tube of borosilicate glass, the lower end of which has been drawn out to a fine tip, with an 80-cm-long bed of medicinal active dry carbon (e.g., Carbopuron, Degussa, Frankfurt, Germany). Fix the carbon between rigid layers of borosilicate glass wool. Clamp the tube vertically to a support. Connect a 3-L suction flask to the lower end of the tube, connect the flask to the vacuum line, tap the tube for 3–5 min with a piece of wood to obtain a tight bed, and pour amyl acetate on the carbon. Close off the connection to the vacuum line after about 30 min. The amyl acetate passes now downward with the aid of the remaining vacuum in the flask. Release the vacuum and disconnect the flask when the first drops of amyl acetate have passed through the column. Place a clean brown glass bottle under the column to collect the eluate. Place a full bottle of amyl acetate upside down into the upper opening of the column.

The amyl acetate will now pass through the column by gravity at a rate of about 200 mL every 24 h. About 6 L of amyl acetate can be purified with one column filling. This is sufficient for at least 3000 analyses.

Use 500-mL dispenser bottles for the reagents, glass bottles for the digestion solution and for the amyl acetate, plastic bottles for the molybdate solution. Only dispensers with high chemical resistance can be used for the digestion solution. The dispenser for amyl acetate must have a high volumetric accuracy.

PROCEDURE: ULTRAMICRO AND MICRO

Place the sample—up to 40 mg of nonrefractory material (e.g., sucrose) or up to 10 mg of refractory material (e.g., polyethylene)—weighed in a silver capsule, Lüdi 76 1308 82, ~8 mg Ag, or in a platinum boat, into the digestion flask. Add 0.50 mL of ultramicro digestion solution and place the flask into the heating block at 250°C. After about 90 min check that the solution boils smoothly without bubbles, which indicates that no more nitric or perchloric acid is present. If it does not, continue the heating. If it does, take out the flask, let it cool, or cool it and add 2.0 mL of the sodium molybdate solution. Mix and add 1.50 mL of amyl acetate. Stopper and shake violently for 10 min. Spin in a centrifuge at 2000–3000 rpm for 1 min. Transfer amyl acetate into a quartz cuvette—conveniently with a disposable pasteur pipette—and measure at 308.5 nm using pure amyl acetate as the reference.

PROCEDURE: TRACE

Place the sample—up to 300 mg of nonrefractory material (e.g., sucrose) or up to 75 mg of refractory material (e.g., polyethylene)—

into the digestion flask. Add 1.0 mL of trace digestion solution and place the flask into the heating block, which is programmed to 90°C for 8 h, 190° for 8 h, and 250° for 8 h. After finished digestion let cool, add 2.0 mL of sodium molybdate solution, mix, add amyl acetate, and extract and measure as described in the immediately preceding section.

SPECTROPHOTOMETRY

Figure 41 shows a calibration curve obtained in routine work with the described ultramicro procedure. No blank was subtracted.

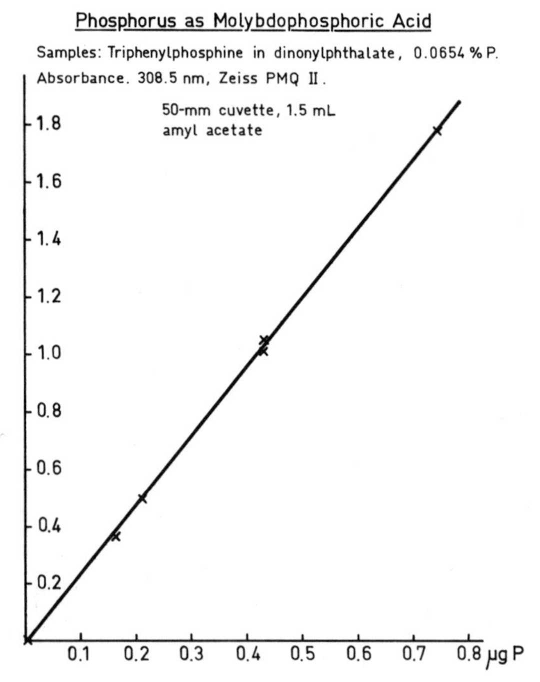

FIGURE 41. Calibration curve for spectrophotometric measurement of molybdophosphoric acid after digestion and extraction according to the described procedure.

Larger amounts of phosphorus can be determined with shorter cuvettes. With the same volume of amyl acetate, up to 40 μg of phosphorus can conveniently be measured with 1-mm cuvettes. Larger amounts can be determined by extraction with larger volumes of amyl acetate or by measurement at higher wavelengths (Kirsten and Carlsson, 1960).

INTERFERENCES

Tungsten(VI) and vanadium(V) interfere by inhibiting the extraction of molybdophosphoric acid. Arsenate causes a small blank.

CHAPTER 22

DETERMINATION OF ASHES

Dry ashing is most frequently used in the analysis of technical, industrial, and agricultural materials in order to compare the combustion residues of different specimens with each other. In this case large series of samples must be burned and the residues must be weighed under well-specified and easily reproducible conditions. Ashing in air in a muffle furnace is most common. The equipment shown in Figure 42 is very convenient for this purpose. The boats with the samples are placed into the numbered compartments of the tray, and the tray is placed into the cold muffle furnace. Usually, no special precautions against creeping and sputtering are needed with samples below 1 mg. The furnace is switched on with the voltage, which gives the maximum desired temperature; when this temperature is reached the reaction is usually finished, and the tray can be taken out and cooled in a desiccator, and the boats weighed.

Clavier and Pouradier (1958) recommend an ashing procedure

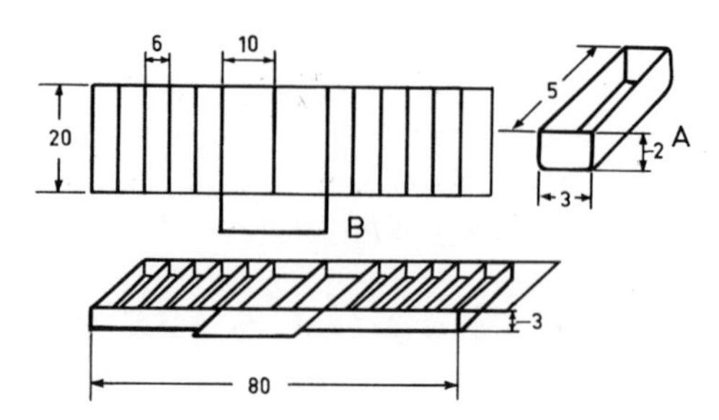

FIGURE 42. Nickel tray and platinum boat for dry ashing. The samples are weighed into the platinum boats (A). Two boats can conveniently be placed into each of the numbered compartments of the nickel tray (B). Nickel plate 0.2 mm, platinum 0.13 mm.

The tray must be superficially oxidized before use to prevent the platinum boats from sticking to the metallic nickel. Dimensions are millimeters.

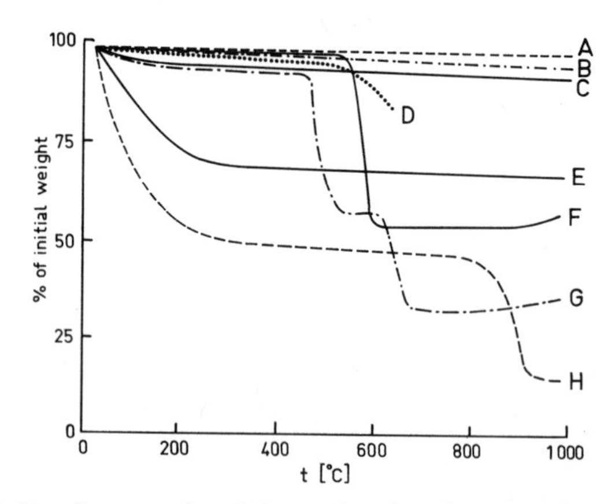

FIGURE 43. Losses of weight on heating for several components of ashes [From Clavier and Pouradier (1958) cited in Bock (1979).] (A) SiO_2, (B) Fe_2O_3, (C) $Ca_3(PO_4)_2$, (D) $CaCl_2$, (E) $Mg_3(PO_4)_2 \cdot 8H_2O$, (F) $CaCO_3$, (G) $Ca(NO_3)_2$, (H) $MgSO_4 \cdot aq$.

TABLE III

Suitable Ashing Temperatures for Different Kinds of Materials[a]

Samples	Ashing temperatures (°C)
Cereals	600
Flour, flour products	550
Starch	800
Jams, fruit juices	525
Coffee, tea	525
Cocoa products	600
Sugar	525
Honey	600
Nuts	525
Spices	550–600
Milk, cream	max. 500
Cheese	550
Gelatine	525
Meat	550
Fuel oils	775
Solid fuels	30 min at 500, then 30–60 min at 815

[a]From Bock (1979).

with a slow heating to 850°C for most organic materials, deduced from their experiences, which are reported in Figure 43.

Suitable ashing temperatures for gram samples of different materials proposed by different authors are reported in Table III.

Several common components of ashes lose weight at certain temperatures. Curves of such losses are shown in Figure 43. In such cases the ashing must be carried through to a temperature where there is a plateau in the curve.

When small, precious samples are to be analyzed for scientific purposes, it is most suitable to carry out the ashing as a thermogravimetric analysis. The thermogravimetric curve provides more and better information than an ordinary ashing procedure.

REFERENCES

Belcher, R. (1966). "Submicro Methods of Organic Analysis,"p. 62. Elsevier, Amsterdam.

Binkowski, J., and Gizinsky, S. (1977). *Mikrochim. Acta* 1977 II, 487–494.

Bishop, E., and Dhaneshwar, R. G. (1962). *Analyst (London)* **87,** 845.

Bock, R. (1979). "Decomposition Methods in Analytical Chemistry," pp. 124, 125. Textbook Co., London.

Clavier, A., and Pouradier, J. (1958). *Chim. Anal.* **40,** 114–117.

Colombo, B., and Giazzi, G. (1982). *Am. Lab. (Fairfield, Conn.)* **14** (July), 38–45.

Colombo, B., Giazzi, G., and Pella, E. (1979). *Anal. Chem.* **51,** 2112–2116

Division of Analytical Chemistry, Commission on Microchemical Techniques (1960, 1961, 1962). *Pure Appl Chem* **1,** 143–145; **3,** 513–515; **7,** 707–709.

Dugan, G. (1974). U.S. Patent 3,838,969.

Dugan, G. (1977). *Anal. Lett.* **10,** 639–657.

Gouverneur, P., and Eerbeek, C. D. (1962). *Anal. Chim. Acta* **27,** 303.

Grunbaum, B. W. (1970). *Microchem. J.* **15,** 680–684.

Grunbaum, B. W., and Kirk, P. L. (1955). *Anal. Chem.* **27,** 333.

Haberli, E. (1973). *Mikrochim. Acta* 1973, 597–606.

Haraldsson, L. (1980). Dept. of Chemistry, Univ. of Lund, Lund, Sweden, personal communication.

Howarth, C. J. (1977). *In* "Instrumental Organic Elemental Analysis" (R. Belcher, ed.), p. 108. Academic Press, London.
Iyengar, G. V. (1976). *Radiochem. Radioanal. Lett.* **24,** 35–42.
Iyengar, G. V., and Kasperek, K. (1977). *J. Radioanal. Chem.* **39,** 301–316.
Kirsten, W. J. (1950). *Mikrochim. Acta* **35,** 1–2.
Kirsten, W. J. (1952). *Mikrochim. Acta* **40,** 121–137.
Kirsten, W. J. (1957). *Anal. Chem.* **29,** 1084–1089.
Kirsten, J. W. (1959). *Proc. Int. Symp. Microchem. 1958,* pp. 132–140. Pergamon, Oxford.
Kirsten, W. J. (1961). *Z. Anal. Chem.* **181,** 1–22.
Kirsten, W. J. (1962). *Proc. Int. Symp. Microchem. Tech., 1961,* pp. 479–493. Wiley (Interscience), New York.
Kirsten, W. J. (1963). *Microchem. J.* **7,** 34–40.
Kirsten, W. J. (1966). *Mikrochim. Acta* 1966, 105–118.
Kirsten, W. J. (1967). *Microchem. J.* **12,** 307–320.
Kirsten, W. J. (1976a). *Anal. Chem.* **48,** 84–87.
Kirsten, W. J. (1976b). *Mikrochim. Acta* 1976 II, 299–310.
Kirsten, W. J. (1977). *Microchem. J.* **22,** 60–64.
Kirsten, W. J. (1978a). *Anal. Chim. Acta* **100,** 279–288.
Kirsten, W. J. (1978b). *Mikrochim. Acta* 1978 II, 403–409.
Kirsten, W. J. (1979a). *Anal. Chem.* **51,** 1173–1179.
Kirsten, W. J. (1979b). *Anal. Chem.* **51,** 2064.
Kirsten, W. J. (1981). *Book of Abstracts.* Lectures held at Euroanalysis-IV meeting at Helsinki, Finland, p. 189. Association of Finnish Chemical Societies, P. Hesperiankatu 3B 10, SF-00260 Helsinki 26, Finland.
Kirsten, W. J., and Carlsson, M. E. (1960). *Microchem. J.* **4,** 3–31.
Kirsten, W. J., and Hesselius, G. U., (1983). *Microchem. J.* in press.
Kirsten, W. J., and Kirsten, R. M. (1979). *Microchem. J.* **24,** 545–552.
Kirsten, W. J., Shah, Z. H. (1975). *Anal. Chem.* **47,** 184–186.
Kirsten, W. J., Danielsson, B., and Öhrén, E., (1967). *Microchem. J.* **12,** 177–185.
Knapp, G., (1975). *Z. Anal. Chem.* **274,** 271.
Oita, I. J., and Conway, H. S. (1954). *Anal. Chem.* **26,** 600–602.
Pella, E. (1978). Istituto Carlo Erba per Ricerche Terapeutiche, Milano, Italy. Personal communication.
Pella, E., and Andreoni, R. (1976). *Mikrochim. Acta* 1976 II, 175–184.
Pella, E. and Colombo, B. (1972). *Anal. Chem.* **44,** 1563–1571.
Pella, E., and Colombo, B. (1973). *Mikrochim. Acta* 1973, 697–719.
Pella, E. (1982). Istituto Carlo Erba per Ricerche Terapeutiche, Milano, Italy. Personal communication.
Pella, E., and Colombo, B. (1978). *Mikrochim. Acta* 1978 I, 271–286.

Poy, F. (1970). *Chem Rundsch.* **23,** 215–217.
Rees, T. D., Gyllenspets, A. B., and Docherty, A. C. (1971). *Analyst* **96,** 201–208.
Ridenour, W. P., Weatherford, W. D., Jr., and Capell, R. G. (1954). *Ind. Eng. Chem.* **46,** 2376.
Schütze, M. (1939). *Z. Anal. Chem.* **118,** 241.
Sisco, R. C., Cunningham, B. B., and Kirk, P. L. (1941). *J. Biol. Chem.* **139,** 1.
Sisti, G., and Colombo, B. (1978). Ital. Patent Appl. 28744-A78.
Small, H., Stevens, T. S., and Baumann, W. C. (1975). *Anal. Chem.* **47,** 1801–1809.
Smith, F., Jr., McMurtrie, A., and Galbraight, H. (1977). *Microchem. J.* **22,** 45–49.
Stoffel, R. (1972). *Mikrochim. Acta* 1972, 242–246.
Sub-Committee for Microanalytical Standards of the Microchemistry Group of the Society for Analytical Chemistry (1962). *Analyst* **87,** 304–316, 400–410.
Terentev, A. P., Turteltaub, A. M., Bondaevska, E. A., and Domotschkina, L. A. (1963). *Dokl. Akad. Nauk USSR* **148,** 1316.
Tölg, G. (1968). "Chemische Elementaranalyse mit kleinsten Proben," p. 24. Verlag Chemie, G.m.b.H. Weinheim/Bergstr.
Unterzaucher, J. (1940). *Ber. Dtsch. Chem. Ges.* **73,** 391.

INDEX

A

Amyl acetate, purification of, for spectrophotometry, 130, 131
Ashes, determination, 135–137
Ashing, temperatures, 136, 137

B

Blanks, water, 50
Bromine
  determination, 3, 5, 113–120
  trace determination, 3, 121–126
Burettes, syringe-type, 9
Bypass system for oxygen determination, 85, 90

C

Capillary, restricting, 46, 69
Capsules for samples, 22, 51, 58, 86
Carbon, nickelized, preparation, 85
Carbon, hydrogen, and nitrogen determination, 2, 4, 38, 43–51
Carbon, hydrogen, nitrogen, and sulfur determination, 3, 53–60
Carbon, hydrogen, nitrogen, and sulfur or trace sulfur determination, 3, 61–66
Chlorine
  determination, 3, 5, 113–120
  trace determination, 3, 121–126
Chromatographic separation—chemical separation, 36
Chromium oxide, granulated, preparation, 43
Chromium oxide–copper oxide, preparation, 45
Cobalt oxide–silver, preparation, 45
Columns
  for carbon, hydrogen, nitrogen,

Columns (*continued*)
and sulfur or trace sulfur determination, 41, 62
precautions, 42
Combustion tubes
filling, 44, 46, 54, 70
nickel, 44, 46, 47, 80
quartz, 44, 46, 54, 70, 80
durability, 49, 80
Connections, flexible, 63
Copper
metallic, 45, 63, 64, 81, 82
ultrapure, 63, 82
Copper oxide, 55, 57
Copper wire gauze, 55

D

Data processing, 39, 40
Detector
flame photometric, 61, 62
electron capture, 82
thermal conductivity, precautions, 41, 42
Doping, carrier gas
with chloropentane, 83, 87–90
with sulfur dioxide, 56, 64, 88, 89
with water vapor, 48, 50
Drying methods, 17–19
Drying pistols, 17–19
Dynamic methods, 1

E

Elemental analyzer, Carlo Erba, 35–42
Encapsulation of samples, 21–29, 74–76, 87
Ethylene blue, color reaction, 5, 95–97, 99

F

Flow regulators, 46
Fluorine
determination, 3, 5, 101–104, 107–112
simultaneously with sulfur, 105, 106
trace determination, 3, 107–112
in oil, 112

G

Gas permeation tube, *see* Doping, carrier gas, with sulfur dioxide
Gas tanks change, 41
Grain analysis, 51, 78

H

Halogen, *see* specific element
Homogenization, 15
Hot-flask combustion, 5, 114–119
Hydrogen determination, *see* Carbon, hydrogen, and nitrogen determination; Carbon, hydrogen, nitrogen, and sulfur determination; Carbon, hydrogen, nitrogen, and sulfur or trace sulfur determination
Hydrogenation of sulfur, 91, 99
Hygroscopic samples, 23–27, 74

I

Injector for solid samples, 13
Inlet delay time of samples, 48, 81
Iodine
determination, 3, 5, 113–120
trace determination, 3, 121–126

K

Kanthal wire, 54
  heater, 54, 55, 57

L

Loop delay, 48, 55
  for oxygen injection, 47, 55

M

Metal salts, determination of oxygen, 87, 88
Methane formation errors, 50
Mills, 15, 16, 78
Minicomputers, 4
Molecular sieves, 86
Multielement methods, 1, 33, 34

N

Nickel combustion tubes, 4, 44, 47
Nickel chloride interference, 48, 90
Nickel oxide, granulated
  preparation, 45
  sintering, 50, 77
Nitrogen determination, 2, 4, 67–78, *see also* Carbon, hydrogen, and nitrogen determination; Carbon, hydrogen, nitrogen, and sulfur determination; Carbon, hydrogen, nitrogen, and sulfur or trace sulfur determination
  errors through retention, 80, 81

O

Oils
  nitrogen determination, 77, 78
  oxygen determination, 88
Oxygen
  determination, 2, 39, 83–90
  recorder traces, 88

P

Peak sensor, 38, 73
Peaks, interfering, 46, 59
Phosphorus
  determination, 3, 5, 127–133
  trace determination, 3, 6, 127–133
Pipettes, 12
  to contain, 11
  to deliver, 11
  Grunbaum, 8
Pyrolysis tube for oxygen determination, 84

Q

Quality control, 4
Quartz tubing
  durability of, 79, 80
  interference in sulfur determination, 99

R

Reactor tube
  for carbon, hydrogen, nitrogen, and sulfur determination and for carbon, hydrogen, nitrogen, and sulfur or trace sulfur determination, 54, 79, 80
  for nitrogen determination, 70, 79, 80
  for oxygen determination, 84
Reagent solutions, 7

Reduction tubes
  reduction, 50, 60
  temperature, 81
Restriction, *see* Capillary, restricting

S

Sampler, 42
  modified, 72
  precautions, 42
Schöniger-flask combustion, 121–125
Separation
  computed, 47
  system for carbon, hydrogen, nitrogen, and sulfur or trace sulfur determination, 62
Shut down of instruments, 41
Soil analysis, 51
Spatula, 13
Spectrophotometry, 31
  in fluorine determination, 102–104
  in phosphorus determination, 132, 133
  in sulfur determination, 95–97, 99
Spot welder, 27, 28, 76
Standard solutions, 9
Standard substances, 9
Sulfide determination through oxidation with iodate, 100
Sulfur
  determination, 3, 91–100, 105, 107–112, *see also* Carbon, hydrogen, nitrogen, and sulfur determination; Carbon, hydrogen, nitrogen, and sulfur or trace sulfur determination
  determination simultaneously with fluorine, 105, 106
  retention on quartz surfaces, 99
  trace determination, 3, 5, 107–112, *see also* Carbon, hydrogen, nitrogen, and sulfur determination; Carbon, hydrogen, nitrogen, and sulfur or trace sulfur determination
Syringes, 11

T

Tailing, 50
Thermogravimetry, 136, 137
Transfers, 11
Tungsten(VI) oxide, 55, 64, 65

V

Volatile samples, 23–29, 74, 76

W

Water gas reaction errors, 49, 57
Weighing, 21–27